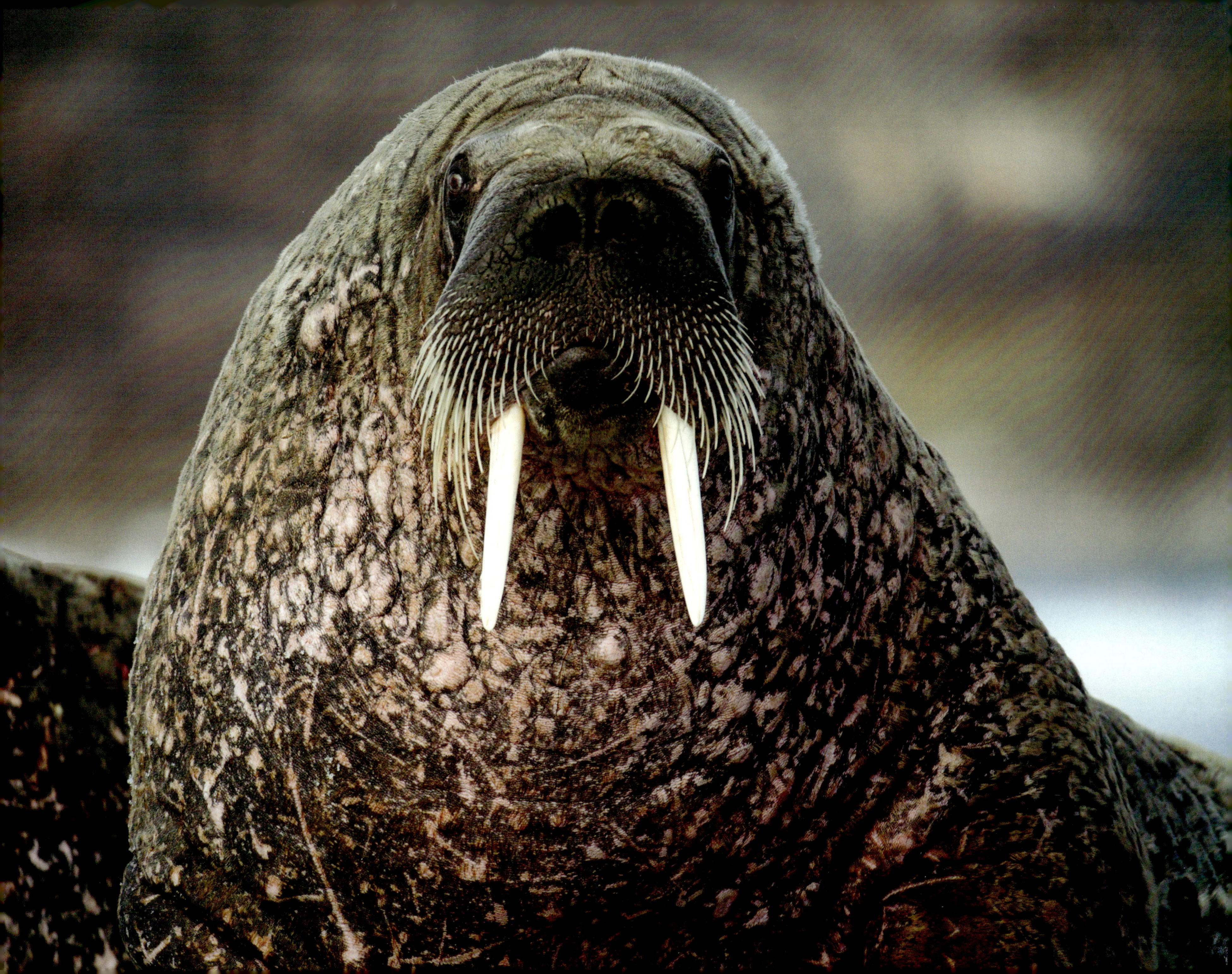

Published by Inhabit Media Inc.
www.inhabitmedia.com

Inhabit Media Inc. (Iqaluit) P.O. Box 11125, Iqaluit, Nunavut, X0A 1H0
(Toronto) 191 Eglinton Avenue East, Suite 310, Toronto, Ontario, M4P 1K1



Editors: Neil Christopher and Kelly Ward
Art Director: Danny Christopher



We acknowledge the support of the Canada Council for the Arts for our publishing program.

This project was made possible in part by the Government of Canada.

ISBN: 978-1-77227-233-8

Printed in Canada

Library and Archives Canada Cataloguing in Publication

Title: Aiviq : life with walruses / photographs by Paul Souders.
Names: Souders, Paul, photographer.
Identifiers: Canadiana 20190045736 | ISBN 9781772272338 (hardcover)
Subjects: LCSH: Walrus—Arctic regions—Pictorial works.
Classification: LCC QL737.P62 S68 2019 | DDC 599.79/9—dc23

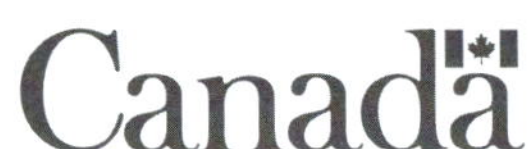

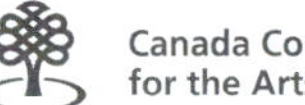

Canada Council for the Arts
Conseil des Arts du Canada

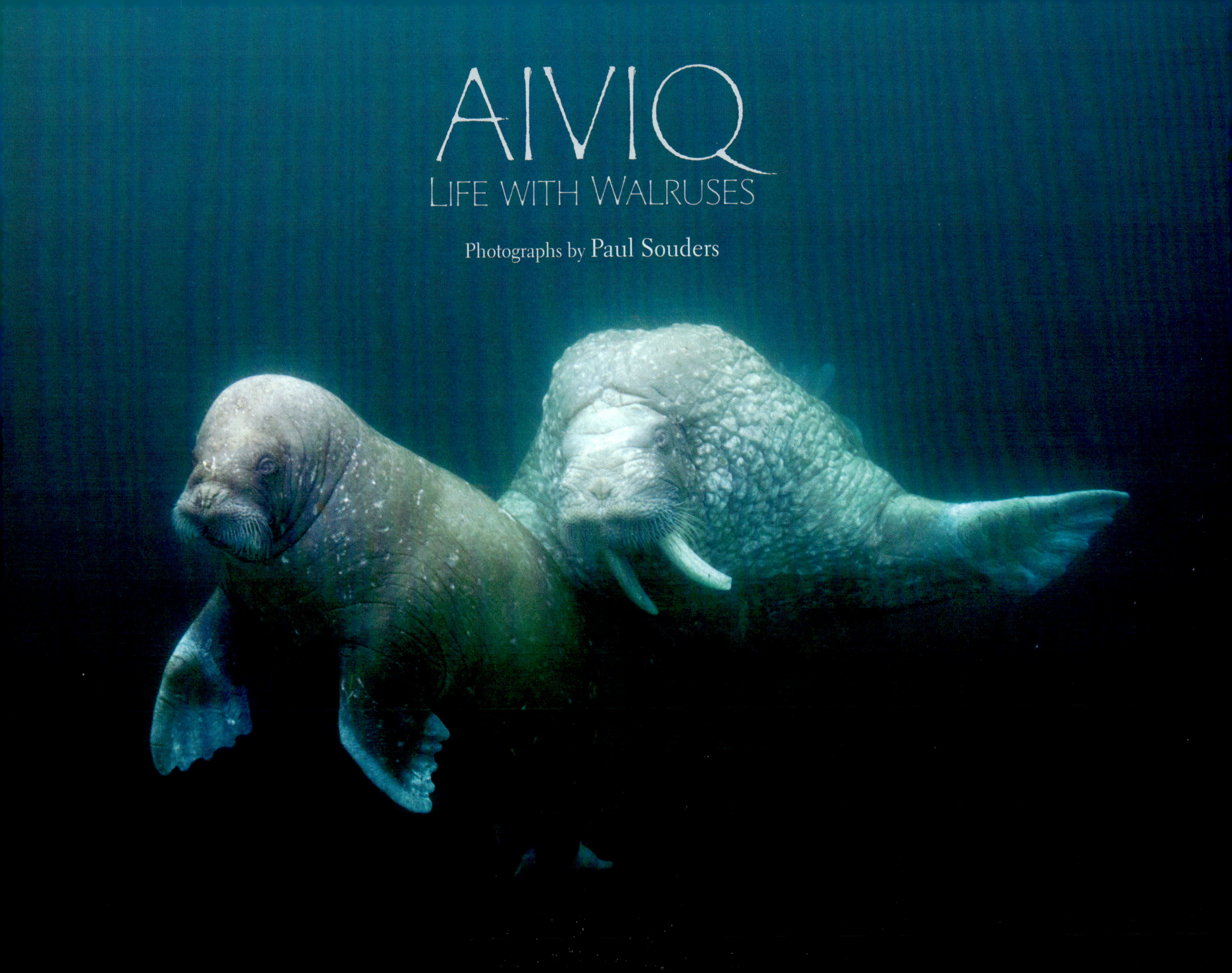

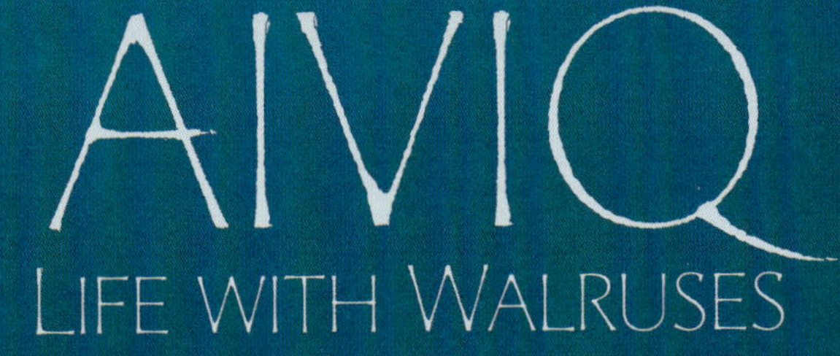


Photographs by Paul Souders

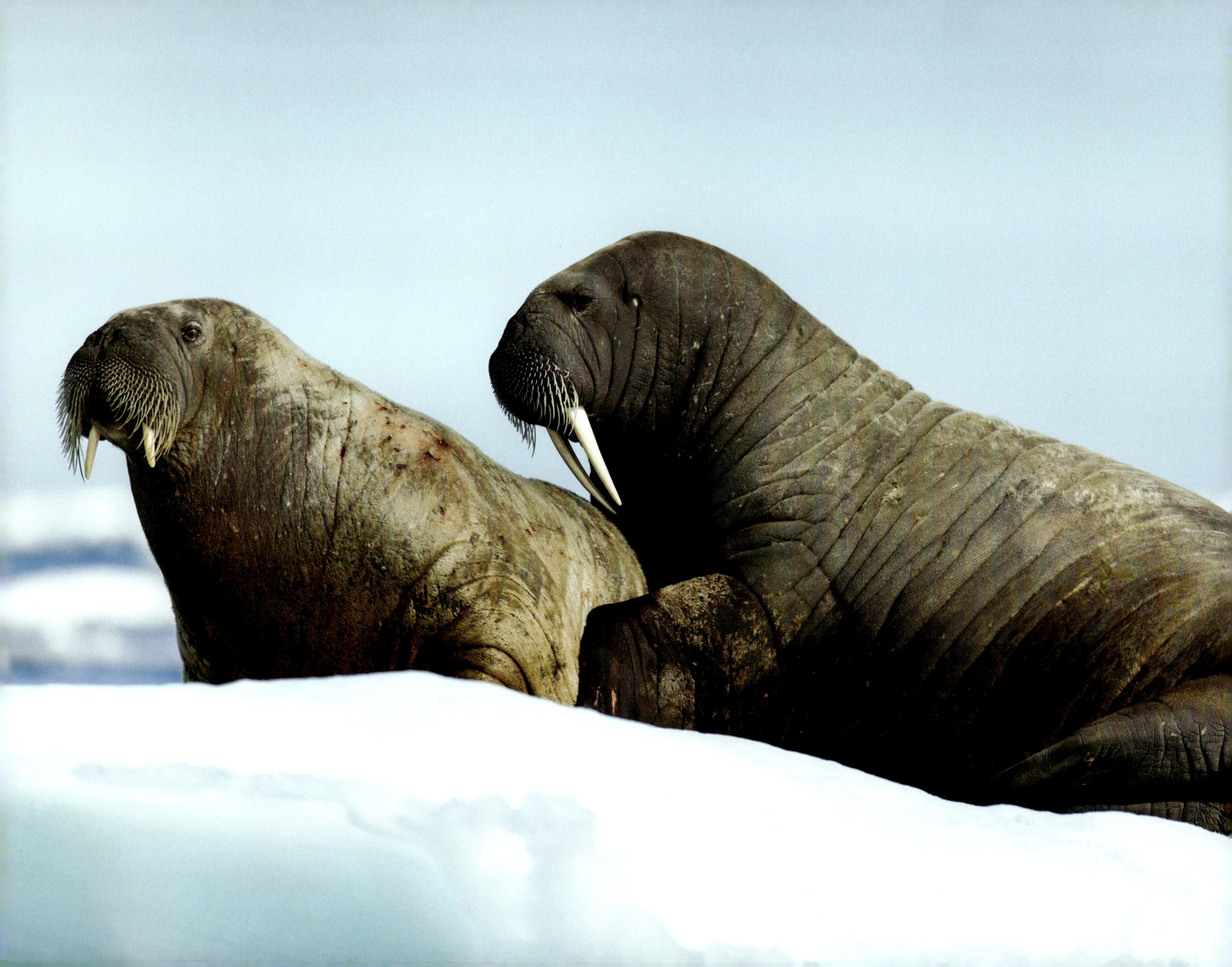

# Foreword

I grew up around Iglulik, Nunavut, where walruses were hunted all year round. My ancestors lived in Avvajja but moved to Iglulik Point during ice break-up to hunt walruses every year. They also went down to the floe edge in mild weather in the wintertime, when the ice is flexible, to hunt walruses.

Every part of the walrus was used—they would use the lining of a gut for a window in our *qarmaq* (a sod house), the meat was eaten by Inuit, and the *igunaq* (fermented, cached meat) was food for dogs. The blubber was used as fat for the *qulliq* (the stone lamp), which provided heat and light, and the teeth and tusks were used for tool making (and they are now sold as carvings). I've also seen whiskers used to make earrings.

My husband, Serapio Ittusardjuat, comes from a long line of walrus hunters. He is now teaching his grandsons to hunt walruses. What I remember most about walrus hunting in Iglulik is when the men came home from preparing the walrus meat for caching and fermenting. They would put the meat inside the thick walrus skin and sew it together, using cartilage to hold it closed at each end. The men would then return to camp in their boat and unload the packages of walrus meat. They carried the packages of meat up to camp, holding them as if in a chain. They walked into camp holding the meat, one man on each end—hunter-meat-hunter-meat-hunter, and so on—until all the meat had been brought to the place where it was to be cached in the gravel. That is one of my most vivid memories of walrus hunting in Iglulik.

*—Monica Ittusardjuat, Iglulik*

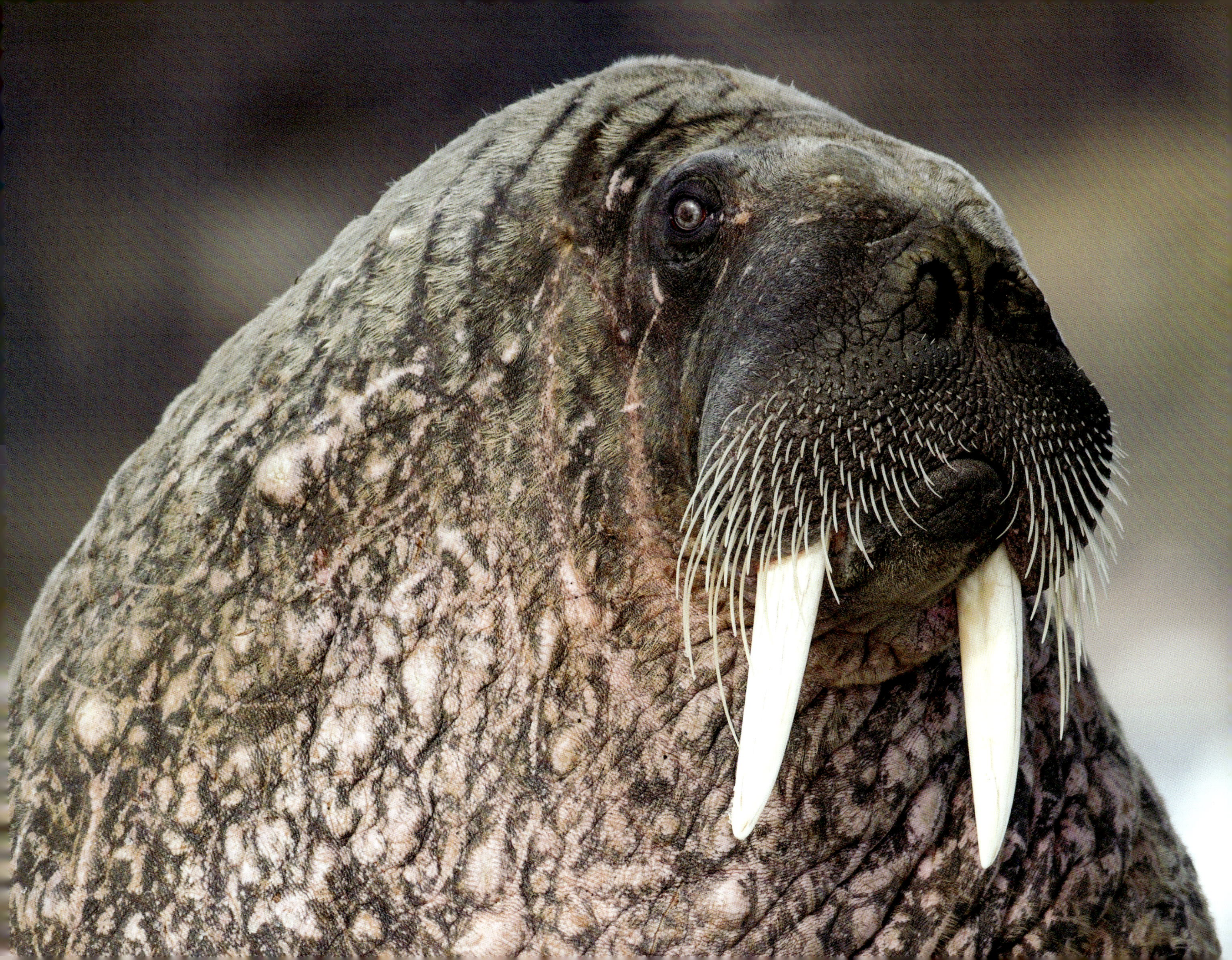

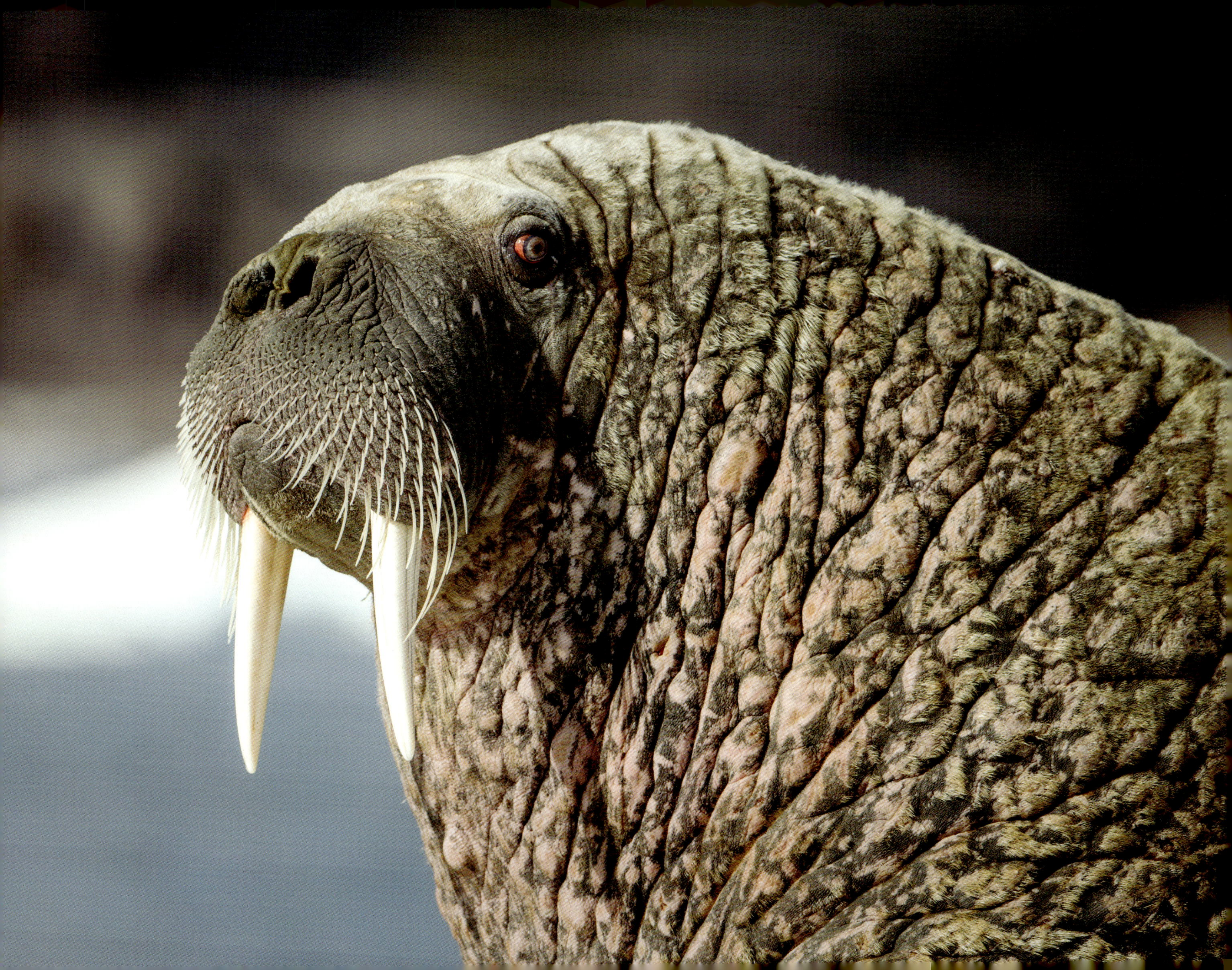

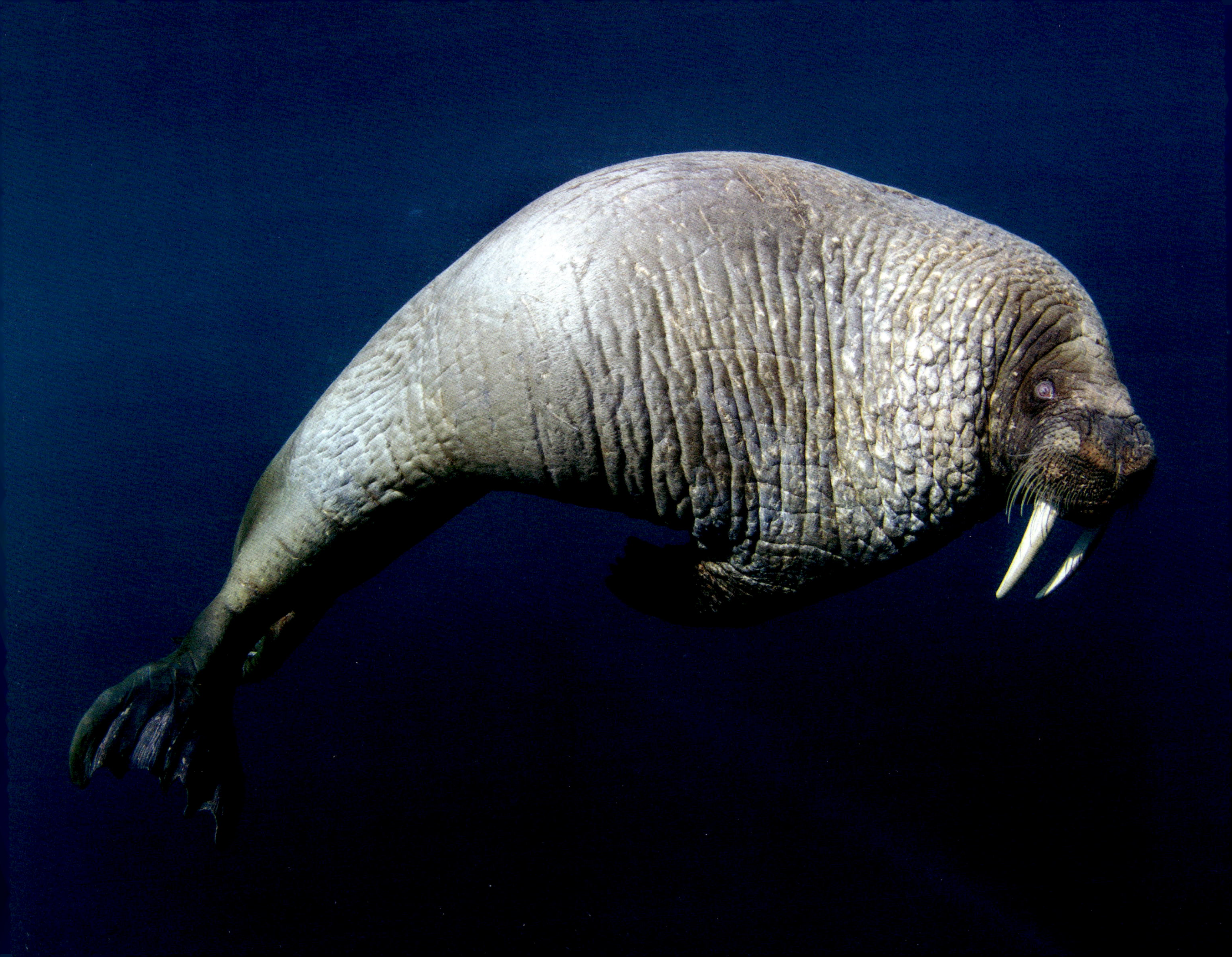

Once I was en route somewhere [when] suddenly my arm was jerked forward, and so quickly that I didn't even have a chance to lay my hand down. I grabbed the oar as I was pulled forward, but in the back of my mind [I was] thinking, "What could be causing the pull?" Apparently, I was dragging a walrus, which somehow had become attached to the oar. Fortunately, I was not injured. Once it stopped moving, I went into deeper waters and dragged the walrus [alongside my boat]. It came up for air, keeping one eye on me. With its loud breath, it sprayed a stream of ice water so forceful that it seemed like the ocean was coming down on me. If the oar had split, there was no way that I could have held on. Fortunately, that didn't happen. The walrus went further away, and I realized that I had been completely soaked. Giita and his older brother began to shoot, but the walrus never did let go. It was frightening.

That was a time that I was caught off guard. After I stopped shaking, I was able to retrieve the catch . . . . We were rewarded with a nice seal for our effort. Once, long ago, I must have had more lives to live, so I have survived to this day.

*—Mialisa Kuniliusie, Pangnirtung*

When I was a teenager, I went walrus hunting with my friends. We came across some walruses. We saw them diving and we were waiting for them to surface. While we were waiting, I looked down into the water and saw a walrus heading straight toward our boat. I quickly rushed out of the way and when the walrus surfaced, we could only see the tusks—it was about to attack our boat. If I hadn't looked down, it would have attacked our boat. We have to be very careful around walruses because they can be dangerous. One thing we Inuit do is when something dangerous like that happens to us, we learn our lesson. We don't repeat anything that once put us in a dangerous situation.

*—Lasaloosie Ishulutaq, Pangnirtung*

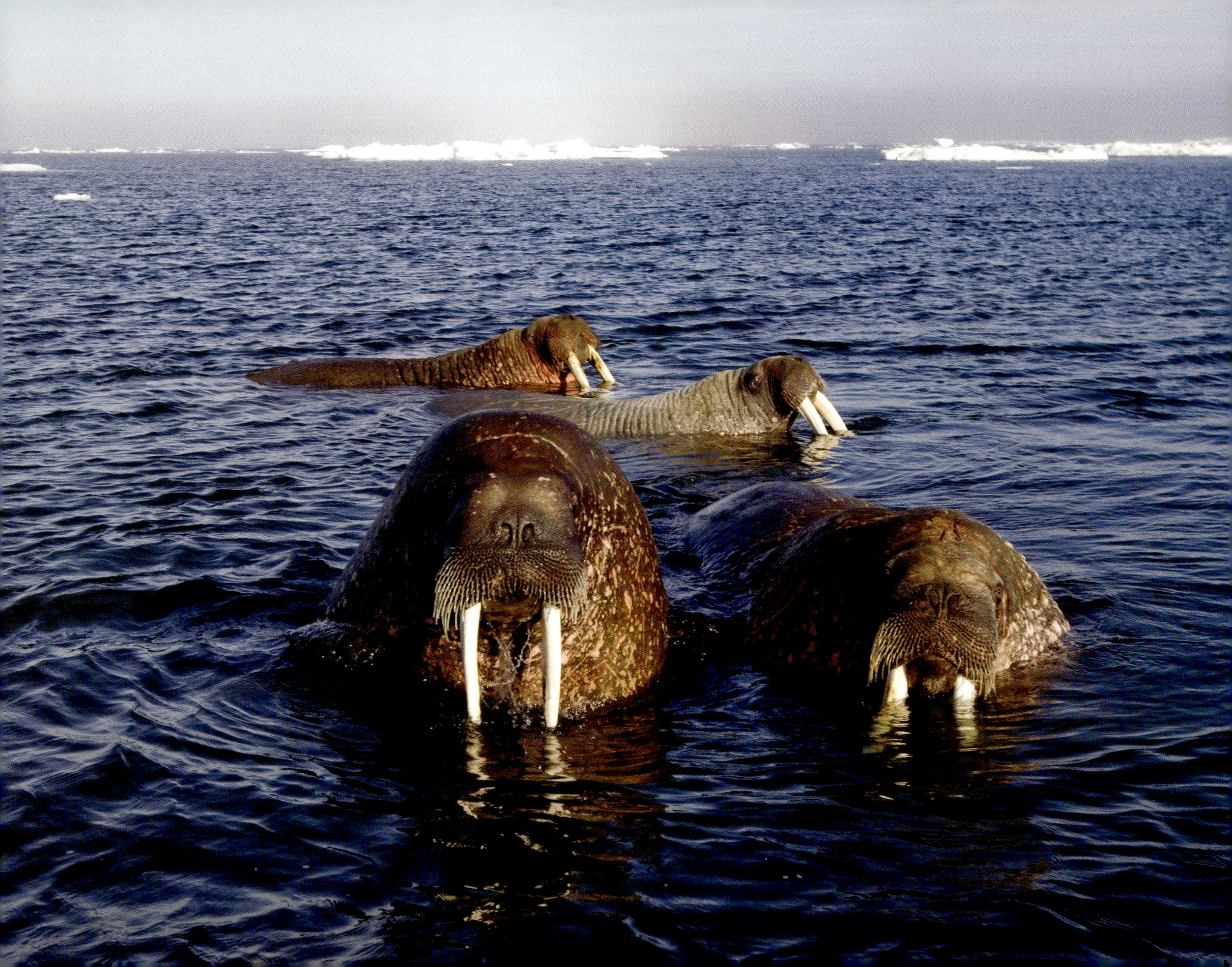

Walruses have about 450 whiskers on their snouts. The whiskers of walruses are highly specialized for detecting prey up to 130 metres underwater, where vision is limited. Walruses find areas with lots of their favourite mollusk prey, such as clams, which are found buried on the ocean floor. Once they find their prey, walruses plow their snouts through the sediment and use their whiskers to detect the size and shape of buried clams. The skin of walruses is thickest just above the whiskers to protect their faces as they search through sediment on the ocean floor. They can use muscles in their faces to erect all of their whiskers at once. Walruses can even move their whiskers individually, using them to open the hard shells of clams to help them eat only the soft inner parts. This allows walruses to avoid eating the hard shells and to leave the shells behind on the ocean floor.

— *Jordan Hoffman, Iqaluit*

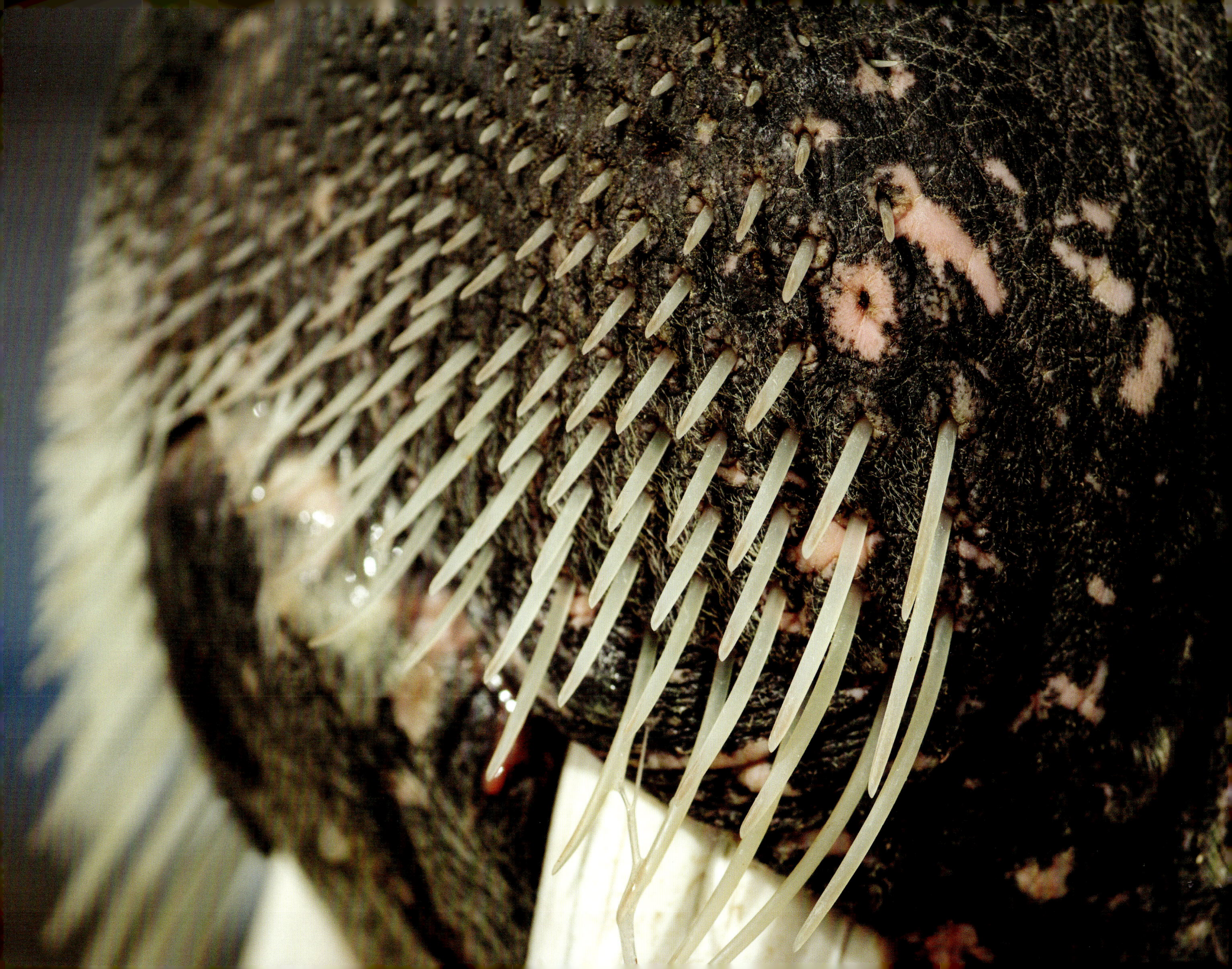

There are different uses for the walrus meat and tusks. The tusks can be used to make a lock for the sled tongue and to make carvings. The meat can feed dogs and it can be fermented and eaten; it's delicious. When a walrus is freshly caught, the flippers are delicious when cooked . . . .

People become very happy when I bring them walrus meat and [they] ask if I'm fermenting meat. When I tell them I am fermenting, they get happy that they'll have some fermented walrus. One walrus can feed many people, and it is eaten by Inuit all over Nunavut.

*—Lasaloosie Ishulutaq, Pangnirtung*

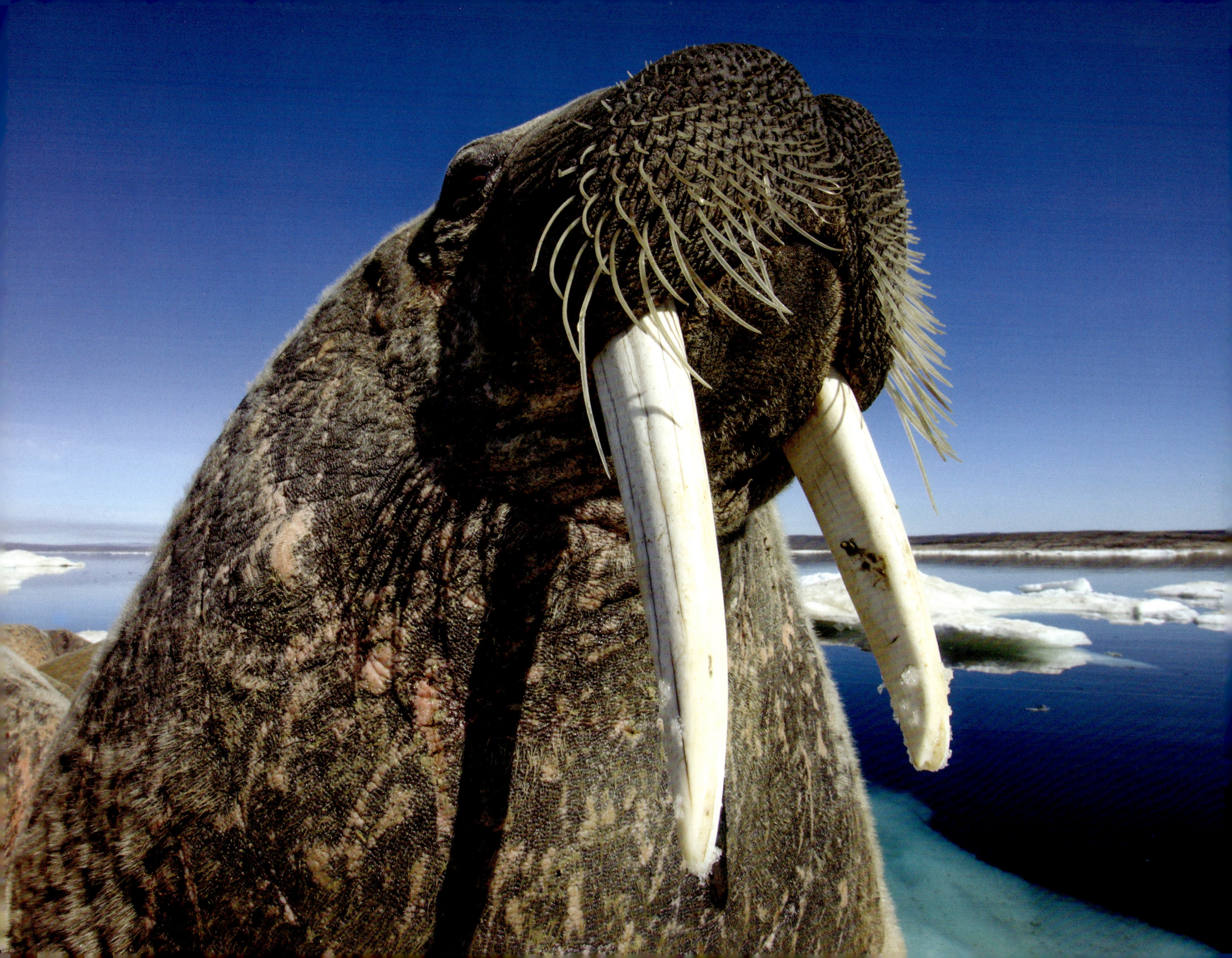

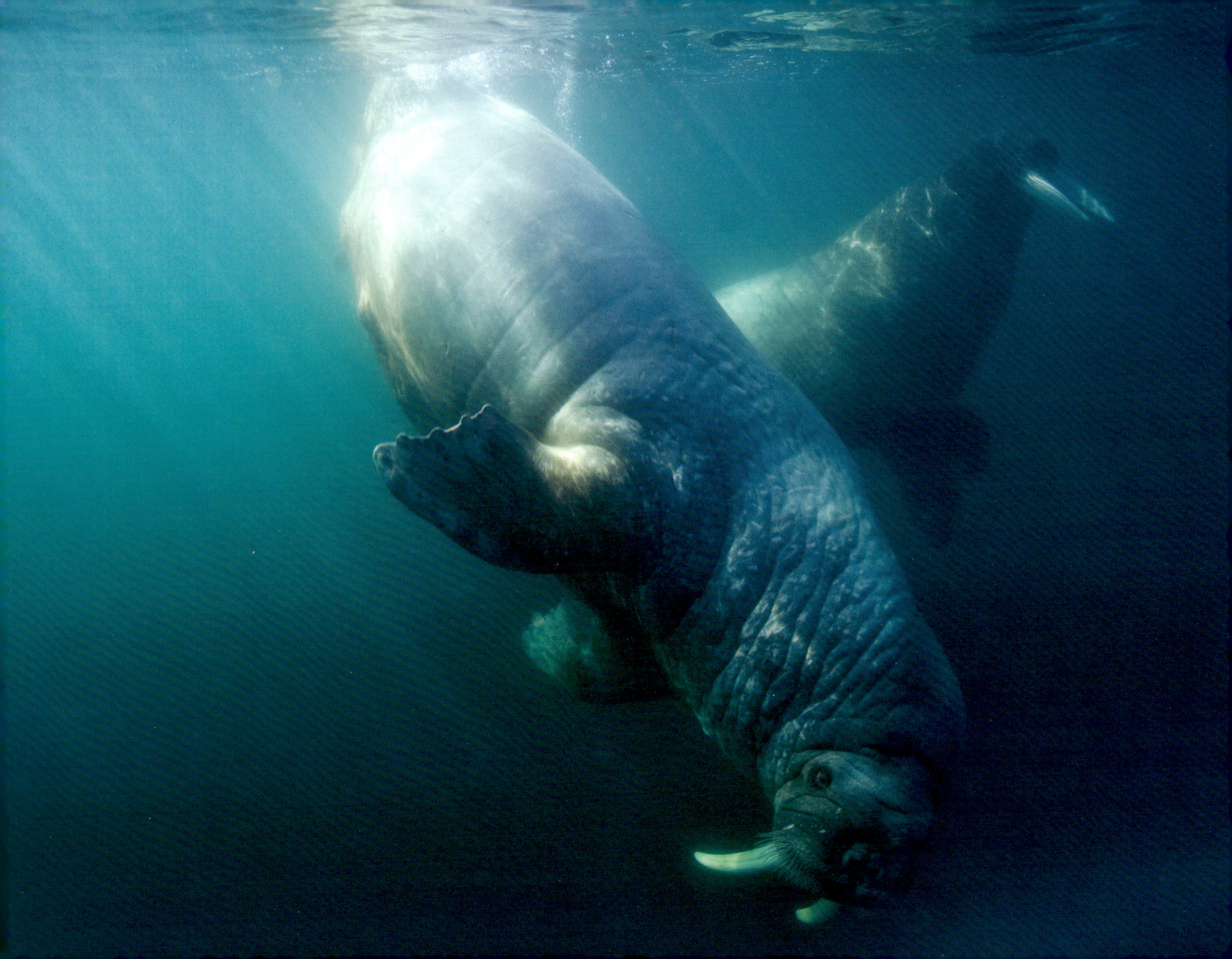

My father and I were hunting, and we went over to see some friends while they were nearby. We saw a walrus with its bum pointing up instead of its head. My father told them not to hunt that walrus because that walrus was aggressive. That's how they show they're aggressive, by having their bum up instead of their head. None of us hunted that walrus. We just continued hunting in the area. A while later, our friends heard knocking sounds on the boat. It turned out that the walrus had been destroying the keel line of the boat. Luckily they were all right and were able to make it home fine.

*—Lasaloosie Ishulutaq, Pangnirtung*

Large male walruses found in the Atlantic Ocean, including the Canadian Arctic, weigh approximately 1,450 kilograms. The largest male recorded weighed almost 1,900 kilograms. These large males can grow up to 3.8 metres long. Walruses found in the Pacific Ocean, including Alaska and the Russian Arctic, are on average even larger than walruses found in Canadian waters. These males can weigh up to 2,000 kilograms.

*—Jordan Hoffman, Iqaluit*

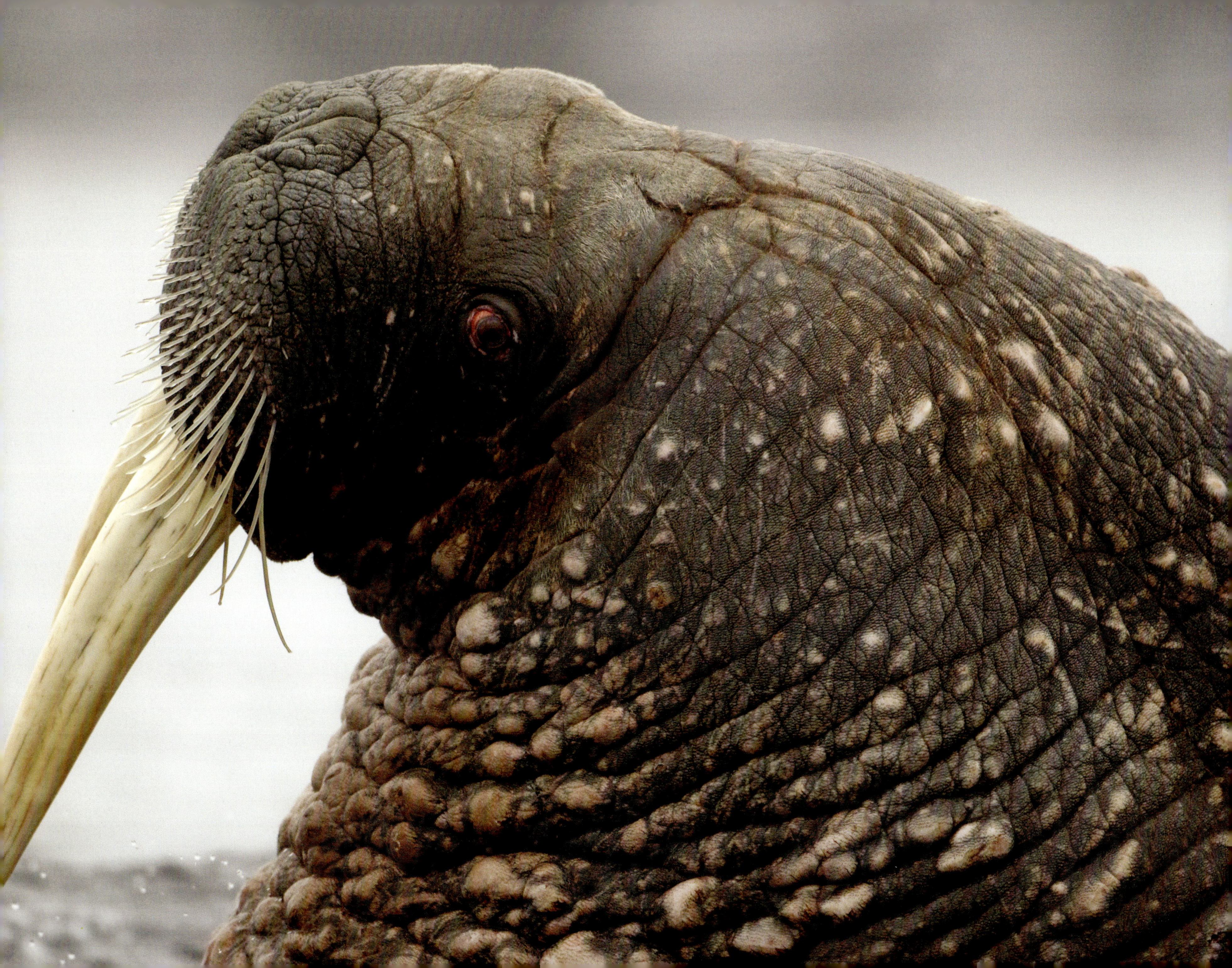

In 2016, I had the opportunity to travel to Coats Island, Nunavut, to assist with thick-billed murre research there. Almost daily, I would see walruses swimming in the water below the cliffs that made up our study plots. A feeling of sheer joy would set in every time I watched them swim together to and from their haulouts. This feeling, however, was nothing compared to seeing them up close.

On a day off, we decided to hike to see an Iceland gull colony and a walrus haulout. When we arrived to the walrus haulout, our bear guide from Coral Harbour, Joe Nakoolak, quickly taught us how to respectfully observe these animals without startling them. Following him, we moved a little closer. The winds must have been in our favour, or maybe it was living by a seabird colony, because I noticed the walruses' size before their smell! A group of astounding males were nearby, and I stood mesmerized by their size, tusks, and movements. Far to the right, I noticed a calf huddled close to its mother—a sight I had really hoped to see! Once the initial awe settled, other senses began to kick in. I had been forewarned about the smell walruses and these haulouts are known for, but what I least expected was the sound. The grunting, the snorting, and most surprisingly of all—the farting! Walrus farts were definitely not something that I had expected to hear or experience, but it's a hilarious memory that will definitely be with me forever.

*—Tianna Burke, Coats Island*

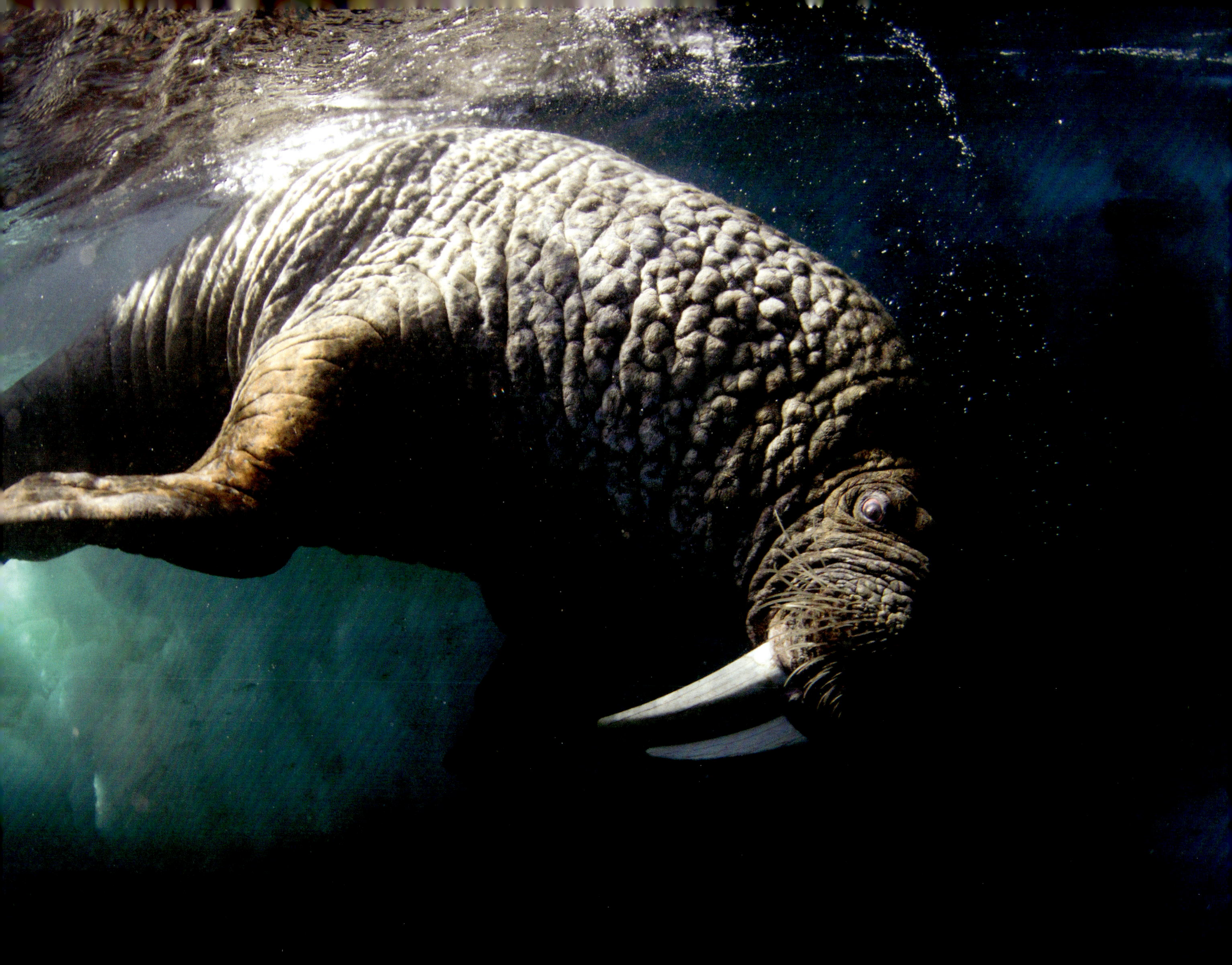

I was a child when my mom and my dad were a part of tourism. We would take *qallunaat* (southerners) out camping . . . . We travelled past Hall Beach to go walrus hunting . . . and went to a little island called Aiviq. It stank.

We were getting closer and the walruses started diving in, but it turned out that we weren't alone approaching the island. There was a polar bear ahead of us. The polar bear went to the island at the same time to look for food. The hunters started off trying to spear the walruses. When the walruses came out of the water, they were huge, and their eyes were red . . . . [There was] a polar bear trying to catch a young walrus. There they were, the father walrus—or it might have been the mother—and the young walrus was in the middle [between the parent and the polar bear] . . . . The polar bear started to attack the young walrus and there was another walrus behind it. Then the [parent] walrus and the polar bear were attacking each other and the young walrus escaped into the water.

The polar bear and the walrus continued to fight. Then the polar bear lost and started walking away [on the ice]. The polar bear didn't give up. It came back to fight again and scratched the walrus around here, on the neck. It started dripping blood. The walrus didn't [want to] lose, so it went underwater with the young walrus. The polar bear also didn't give up . . . . It saw and targeted a walrus that [my stepfather's older brother had] speared and shot . . . . They were trying to shoot the walrus we had speared, and then all of a sudden, they were trying to shoot a polar bear, and some forgot about the walrus! I was just sitting there, watching the chaos. As I watched, some hunters were trying to catch the polar bear, and the others continued to put full effort into trying to pull the walrus out of the water.

*—Corianna Manitok, Iglulik*

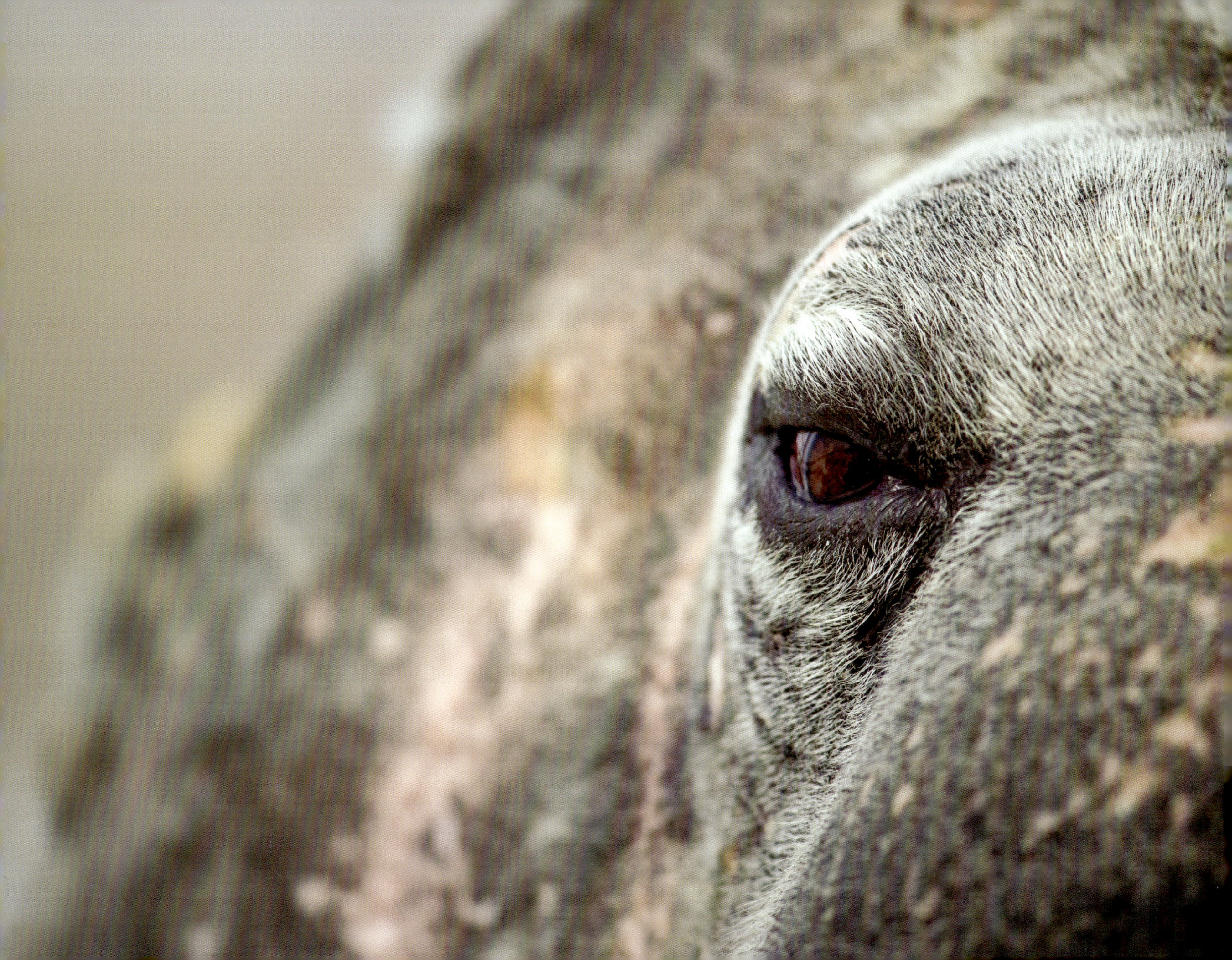

There will always be walrus hunters because this is the Arctic . . . . I teach walrus hunting, everything from getting ready to caching it or bringing it home. The reason for this education is that it's very important to use Inuit traditional knowledge, especially with food. We've always been told not to hunt more than we need, only hunt what you need. We used to harvest so many more walruses back then because we had more dogs to feed. We have fewer dogs now, so we don't hunt as many walruses. We were told that walruses can be dangerous to eat because they can cause illness, but when you really pay attention to animals, you learn more about them. Like when a walrus is sick and can be dangerous to eat, it won't be around other walruses and will tend to be aggressive. When walruses are being aggressive, they do not have their heads up but instead they have their bums up. It would make more sense to be aggressive with the tusks, but they don't show them. Their aggressive ways can be different, so you have to be careful around them. When I see a skinny walrus, I observe it more and assume it is sick.

*—David Irngaut, Iglulik*

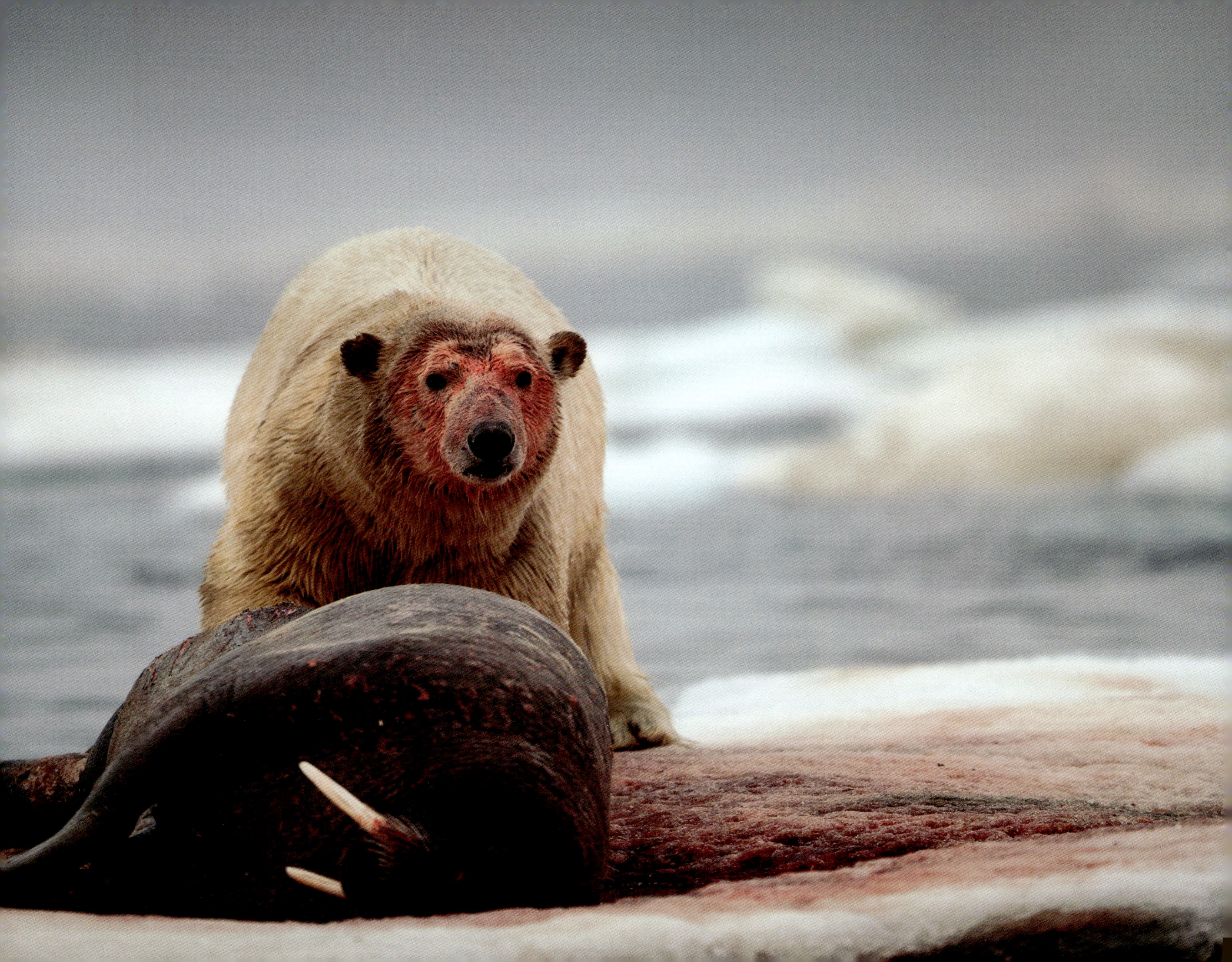

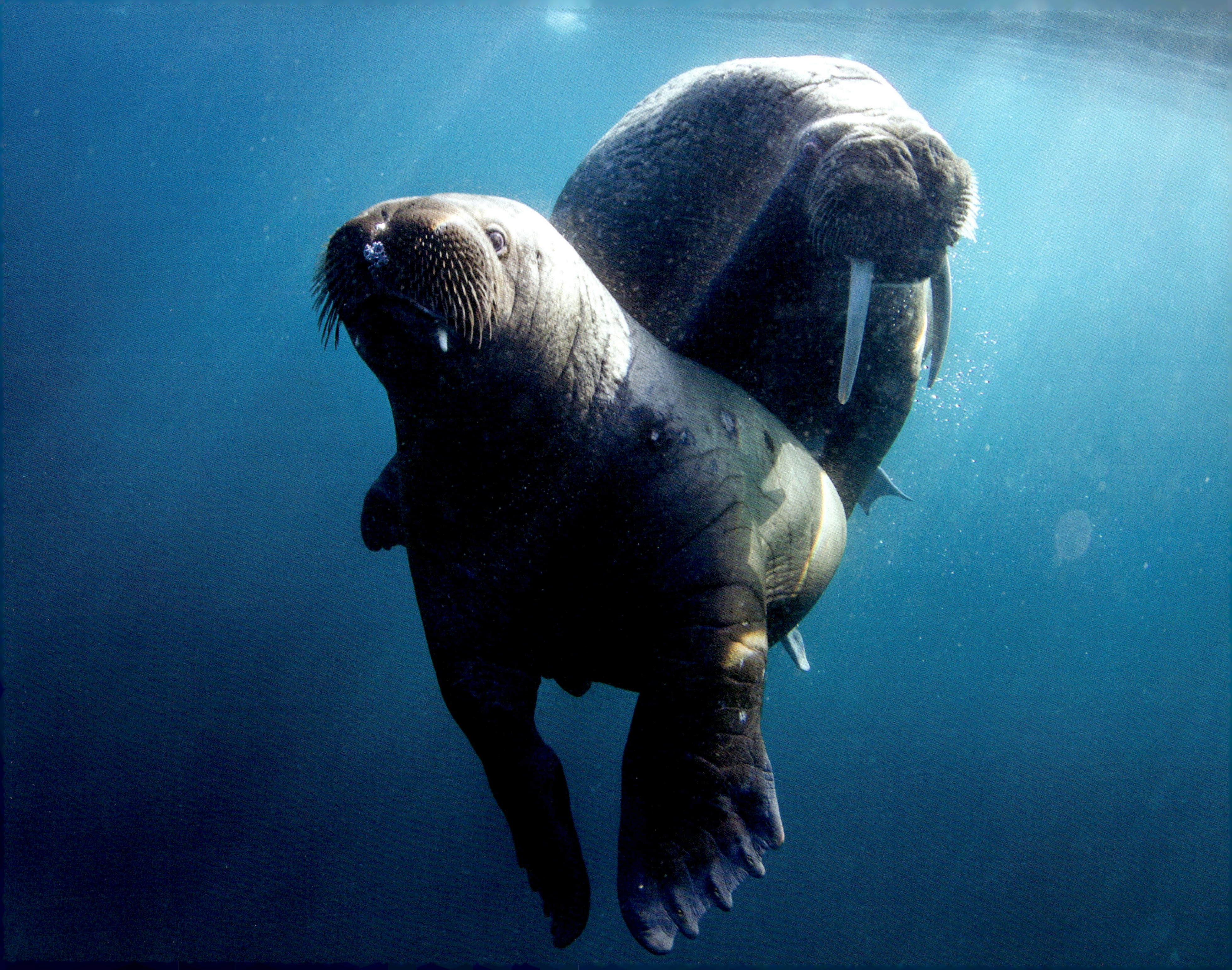

When the ice is good enough to go hunting on in the winter, I enjoy going walrus hunting. We also go walrus hunting in the summertime. That's when we hunt to cache the meat. Mid-July was the best time to cache walrus meat, but now that summer gets warmer, we have to dig deeper to cache the meat. Back then, the meat was mainly to feed the dogs, but today, we want to eat the meat more along with the fat. We like to hunt walruses all year round. They will start coming closer to the floe edge in March. There will be smaller ones with their young, and they will be starting to have babies [around that time]. I enjoy hunting them very much.

*—David Irngaut, Iglulik*

When we sew up walrus meat in its skin to cache the meat, we need to make sure it's full and tight. It will make a good fermented meat that way, with no air to make it go bad. It's better to make it tight and dig deeper to ferment it properly, because you have to do it a specific way to do it right.

It seems like it is only us around here who like walrus meat, but there are Inuit from different communities wanting to order walrus meat. They like it too, and we Inuit really depend on country food. Some of us can't eat food from the stores for too long before we get really sick of it.

*—David Irngaut, Iglulik*

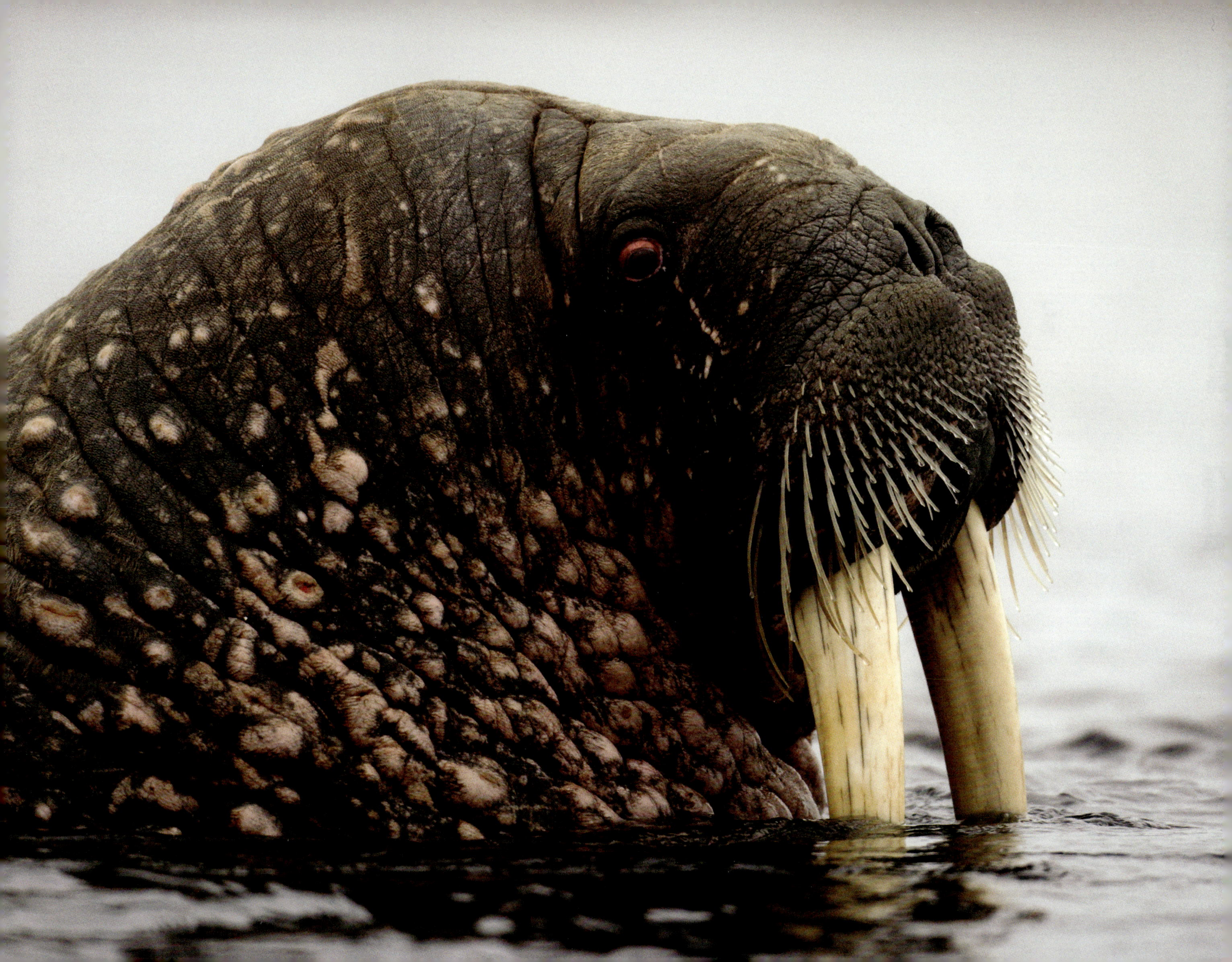

There was once a girl at Ârviaq (Sentry Island) who played with the bones of a walrus, pretending [the walrus] was her husband.

Then it happened that kayaks out at sea were pursued by a walrus, and no one could understand why the walrus kept coming after them. Then it was discovered that there was a girl in the village who played with the bones of a walrus, pretending [the walrus] was her husband.

But the men threw the girl into the sea, out to the walrus, and when it had thus obtained its wife, and its own way, it never again pursued the men who were out hunting.

*—Aqikhivik, traditional story recorded in the Kivalliq Region, 1920s*

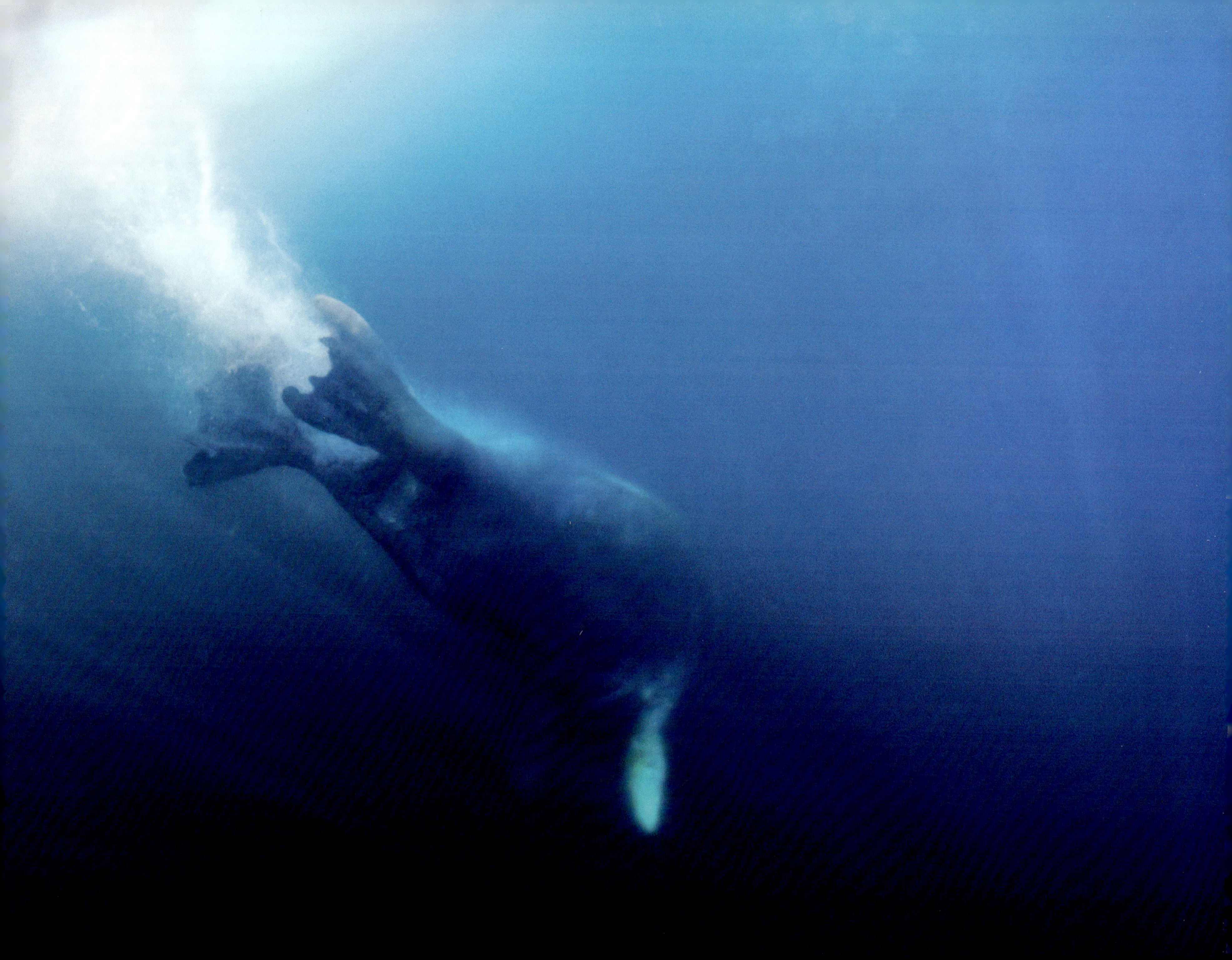

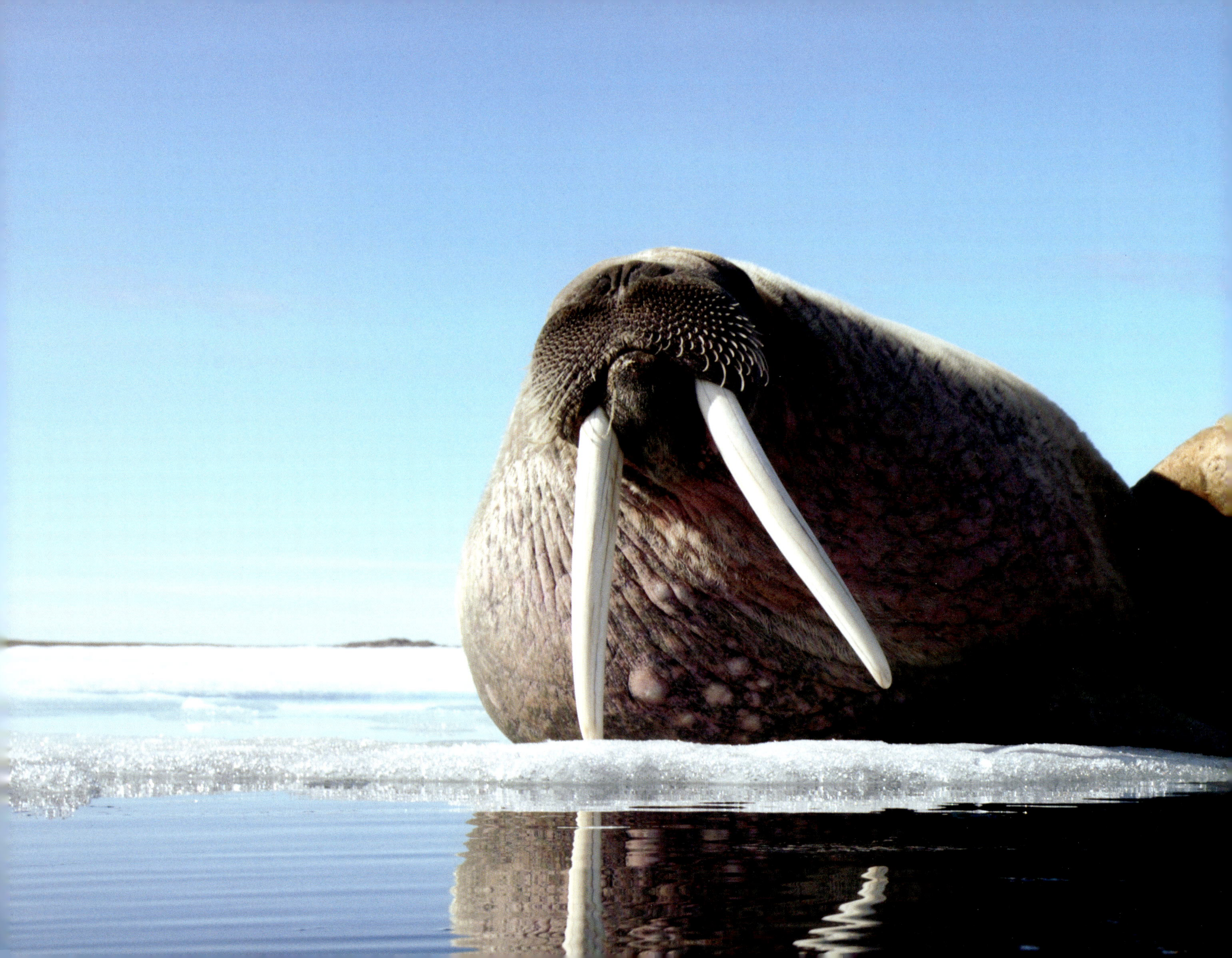

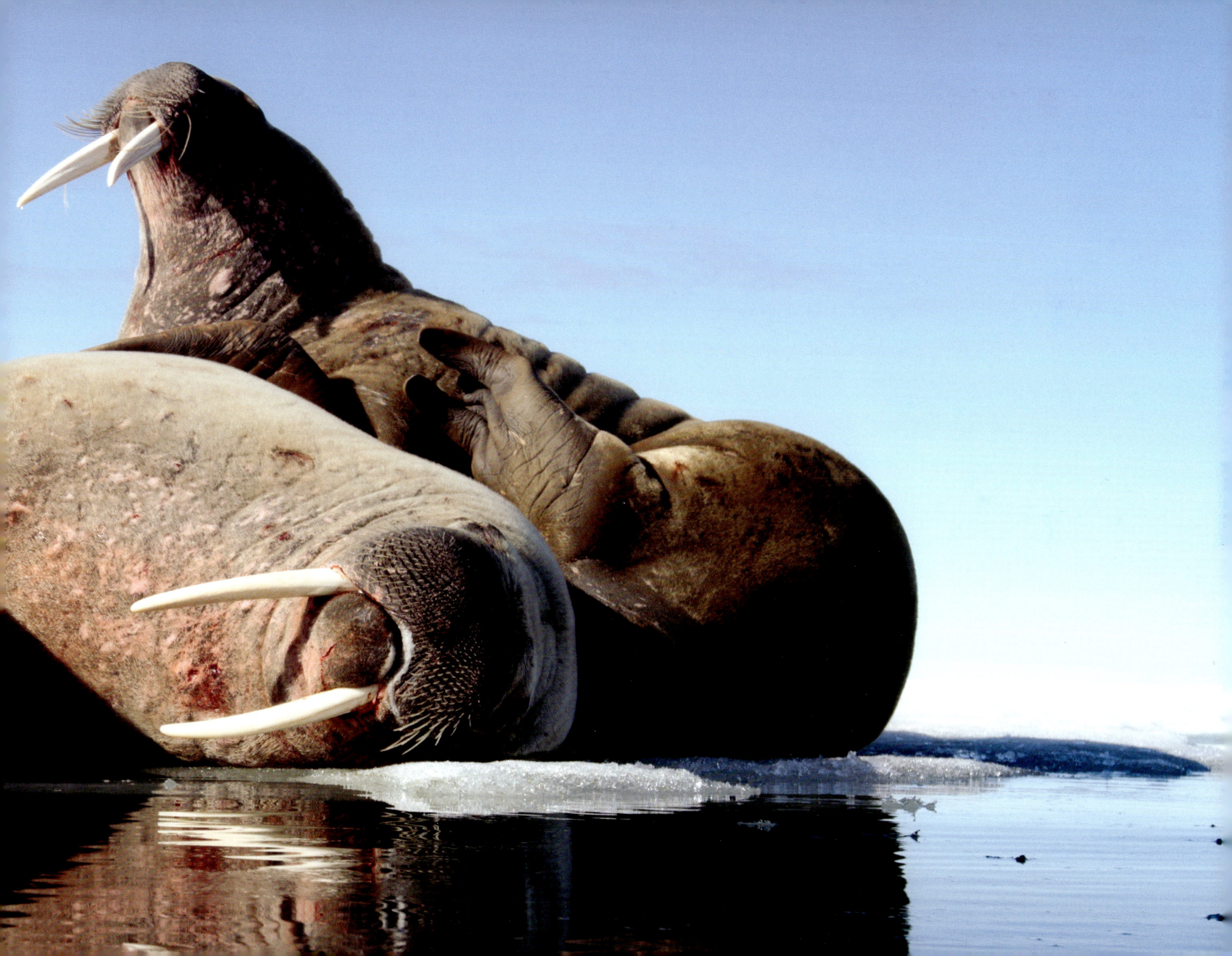

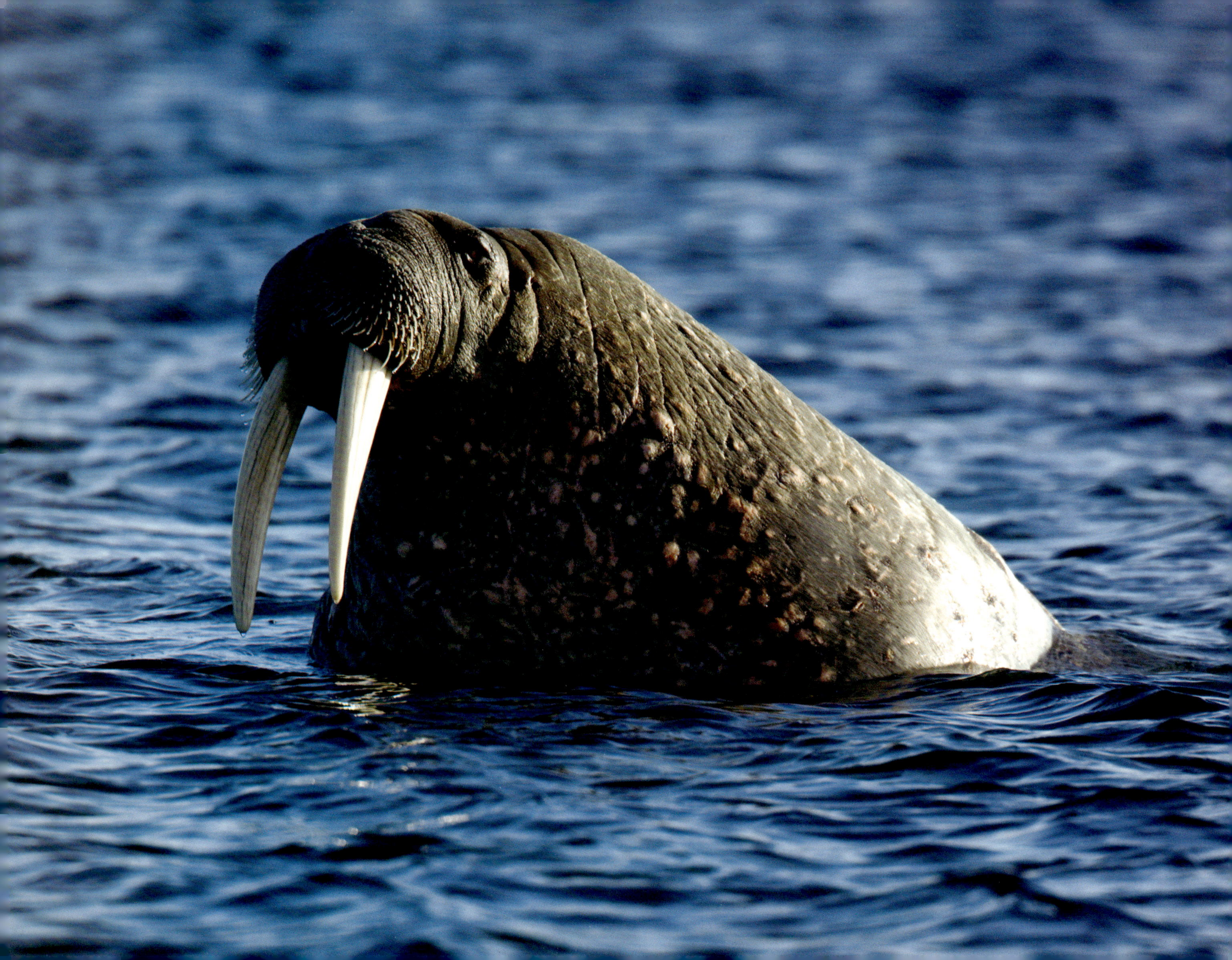

Walruses eat clams and seals. Clams prevent them from getting parasites while seals give them parasites. When we go to the floe edge, we know there are walruses around when there were seals but there's nothing there anymore. We as hunters know this . . . . Walruses make loud sounds; they sound like this, "uuq uuq uuq," and they breathe heavily, which is why seals fear them . . . . Walruses eat seals by hugging the seal from behind. Hunters have lost their seal catch because a walrus ate it.

—*Lasaloosie Ishulutaq, Pangnirtung*

[A traditional story says that] Nillatuarjuk was a small man who farted a lot. While he was out on the ice, there were four walruses basking on the ice. Nillatuarjuk farted on the first two, so that they weren't good to be eaten anymore, but the other two were good to be eaten. This is the reason why walruses are stinky.

*—Lasaloosie Ishulutaq, Pangnirtung*

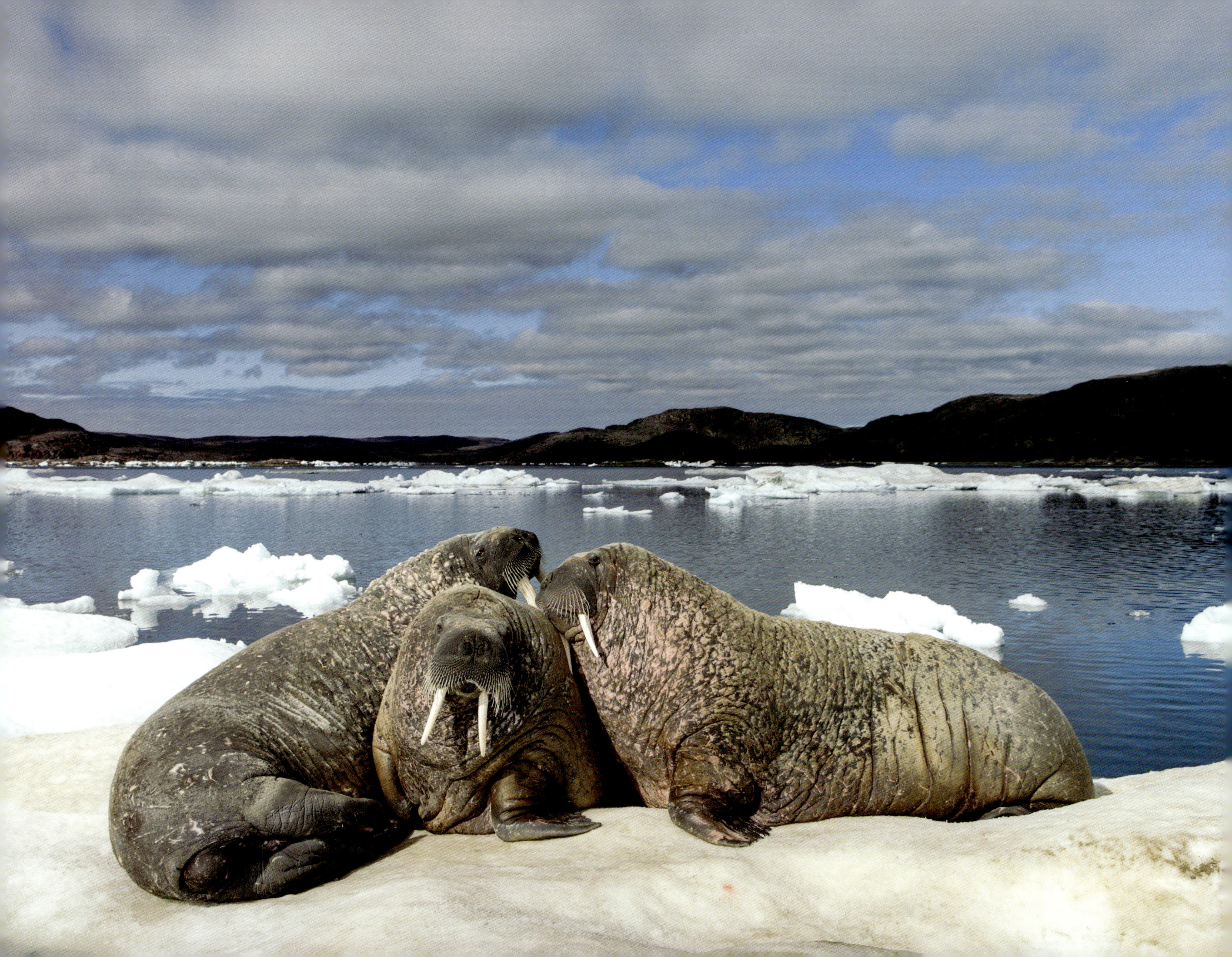

Once upon a time, an old woman who had died was buried, and then a raven came and began to eat her. Her soul entered the body of the raven, and she became a raven . . . . Now the soul went into a walrus, and the walrus had young ones. This walrus became hungry and went down to the bottom of the sea to dig clams, but the clams would not open their shells, and it came up still hungry. It said to the other walruses, "I cannot get anything to eat. The clams refuse to open their shells for me." Then the other walruses said, "When you go to the bottom of the sea, say, 'Eok, eok, eok!'" It did so, and as soon as it said, "Eok!" the clams opened their shells, and it had all it wanted to eat. Soon after this the walrus was caught by a man, and the soul of the woman went into a ground seal, which had young ones.

*—Traditional story recorded on the west coast of Hudson Bay, 1800s*

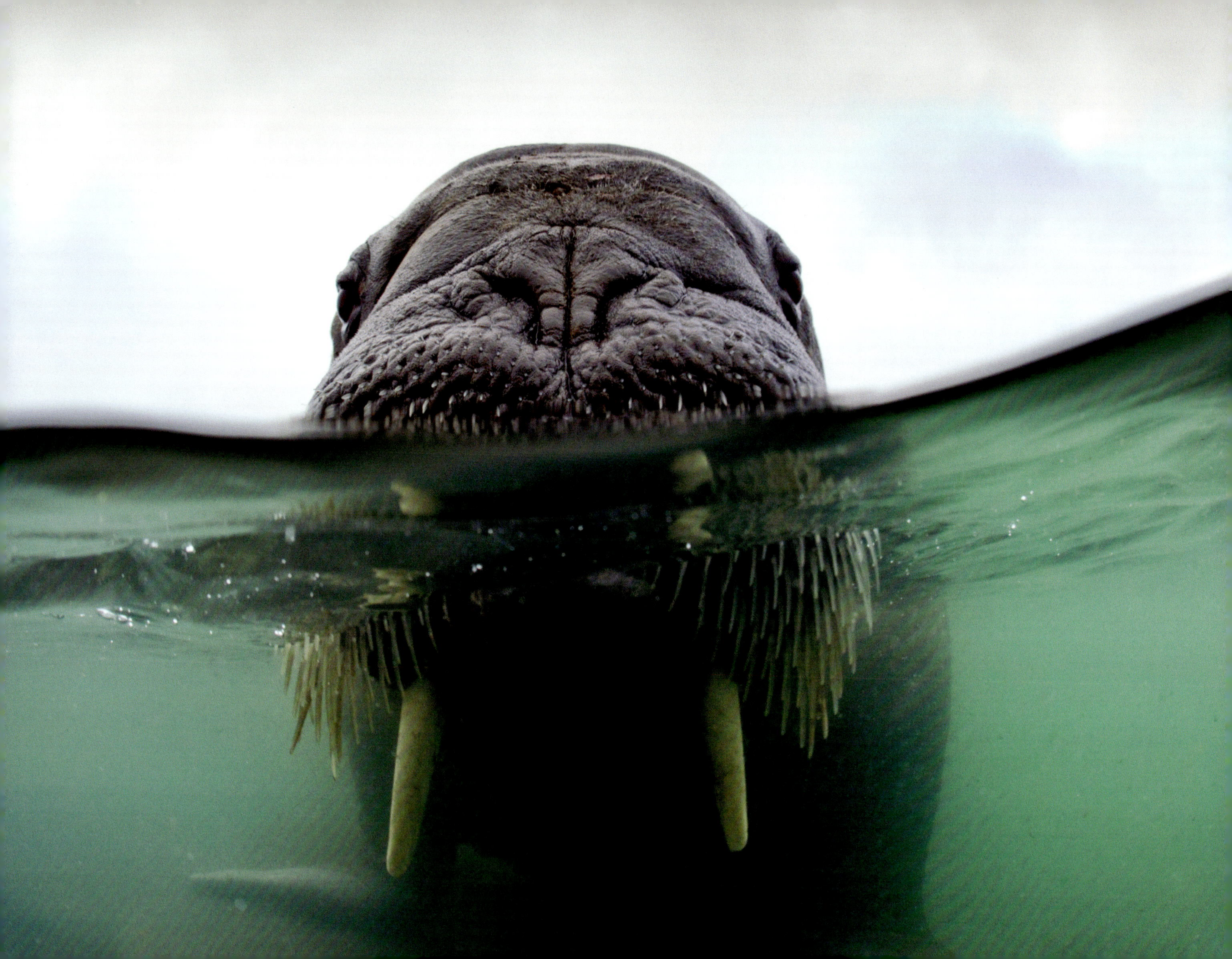

I'd imagined going swimming with a walrus for years, but it took a lot of planning and effort to make it happen. I had to charter a steel-hulled yacht in Svalbard, Norway, six hundred miles north of Europe's edge, and drag hundreds of pounds of scuba and other underwater gear from my home in Seattle through four different airports. Walruses are, by reputation, large, aggressive, and unpredictable. It took an enormous leap of faith to get into the water with them. But sometimes you just have to hope for the best. We'd sailed to one of their haulouts in the Svalbard archipelago, and found a few of them swimming in the water. I quickly put on all of my gear and then, as a last-minute nod to safety, tied an old piece of rope around my waist for the boat skipper to hold on to. And then I slid off the iceberg and into the water.

I could see the walruses circling below me in the depths, and they slowly came up, equal parts wariness and curiosity. They seemed impossibly big, but quite graceful and gentle as they swam towards me. One grew bold and was soon pressing his whiskers and tusks against my camera's glass dome. He even gave it a little tap, just to see if I was paying attention.

Apparently blue-lipped, hyperventilating photographers aren't all that distracting after all, since they soon swam off, taking one last look at me before diving again into the depths.

*—Paul Souders, Svalbard, Norway*

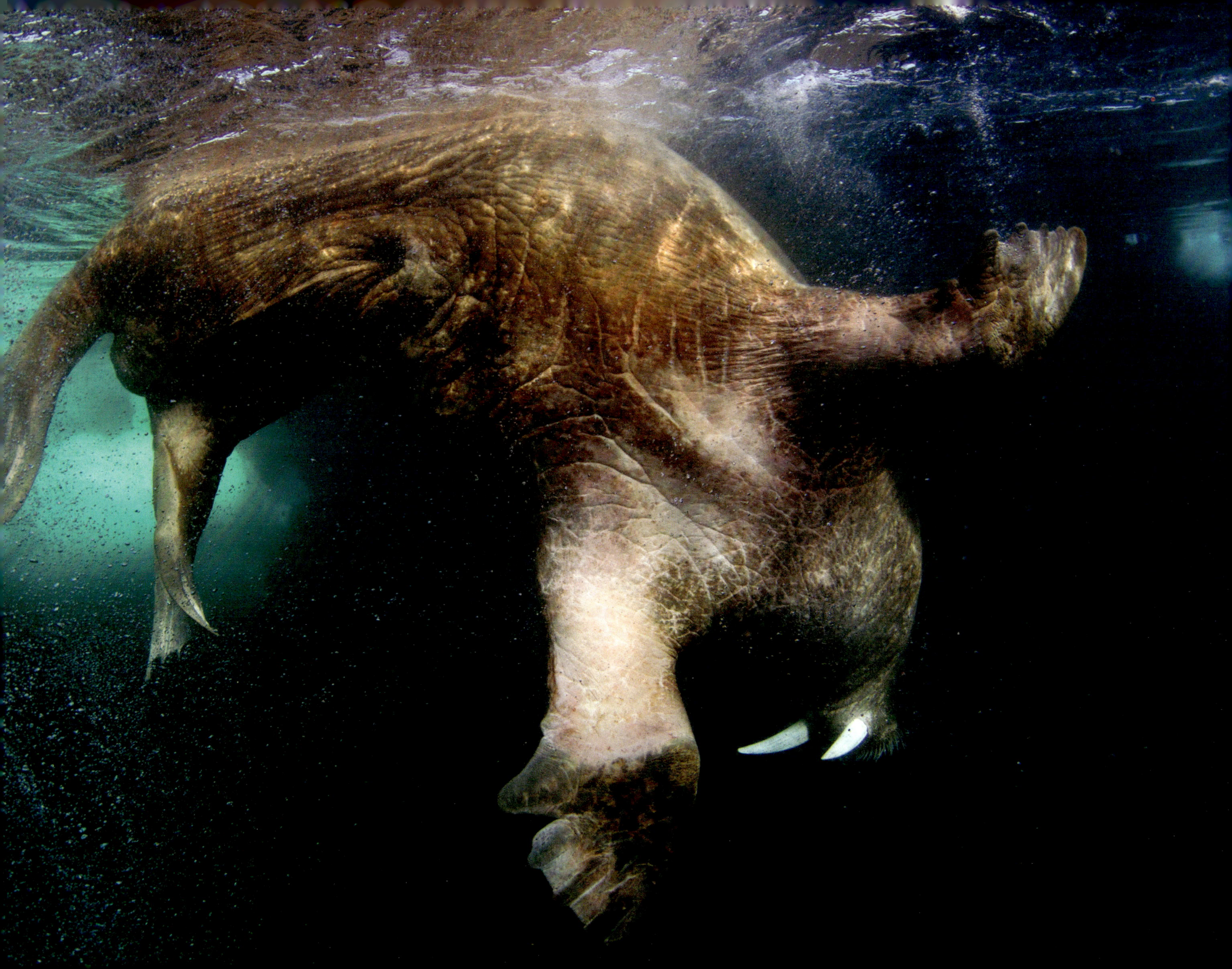

Once upon a time, two boats started from Natsilik down Koukdjuaq, and northward along the shore of Fox Channel. One of the boats returned the same summer. They had taken some walrus hide along, and Piaraq, their leader, had said to them, "Why do you take walrus hide along? You must not take walrus hide when you go caribou hunting. If you do, you will starve." But they did not follow his advice. They found only a few caribou at Natsilik and had to return for want of food. After some time, some other people went down Koukdjuaq, trying to find the other boat's crew, who had gone along the coast of Fox Channel. They fell in with them in the fall and found that they were starving.

*—Recorded in the Cumberland Sound area, 1800s*

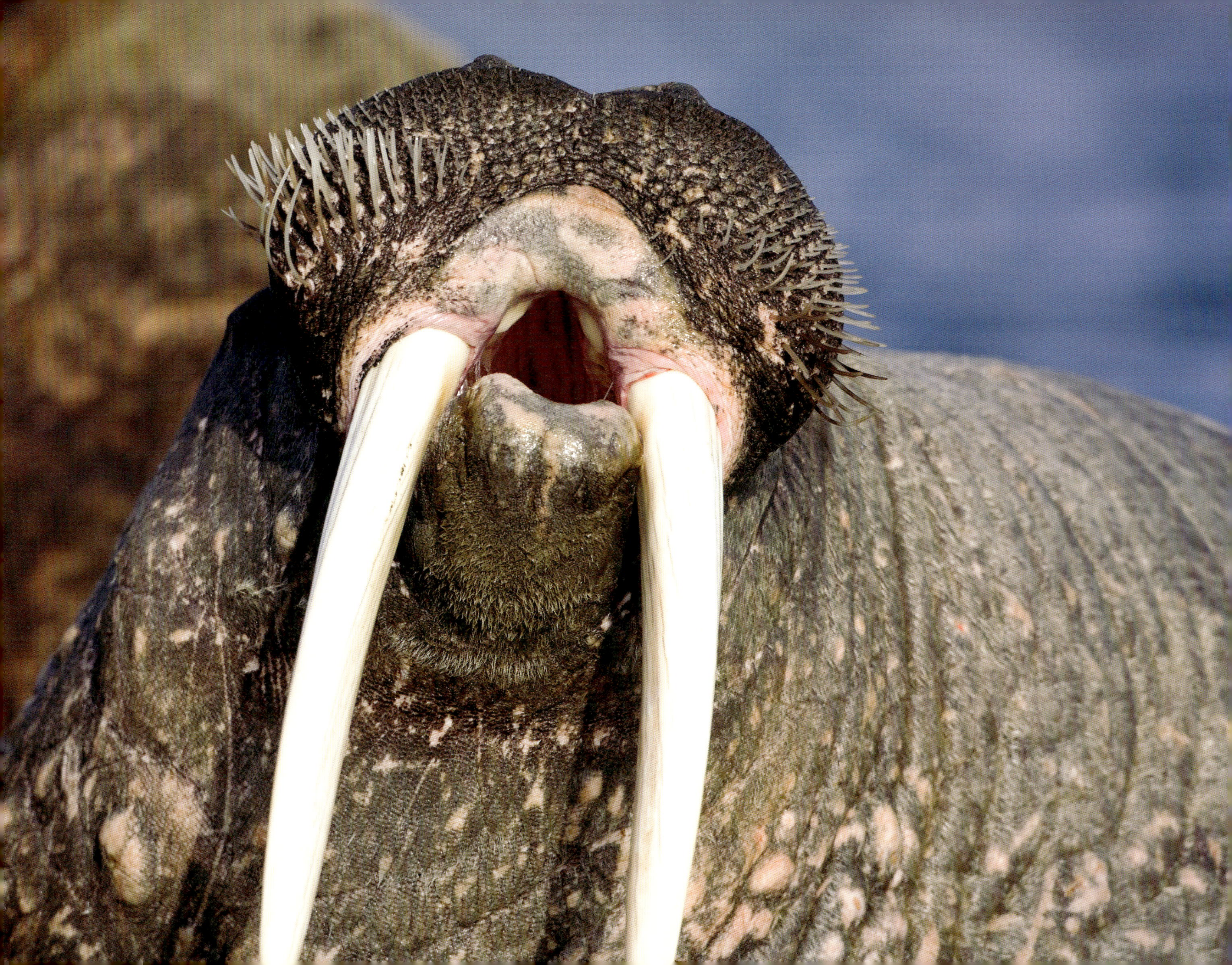

When walruses are basking on ice, a hunter shouldn't whisper or try to be sneaky. Instead they should be loud. When there are not a lot of walruses together in the water, they startle easily. It's easy to lose them, and they come back up for air so far away from where they [first] appeared. While going after them on ice, they can get up high, and they look very intimidating. It's fun while walrus hunting, but we must be careful and observe carefully because walruses can be fierce.

*—Lasaloosie Ishulutaq, Pangnirtung*

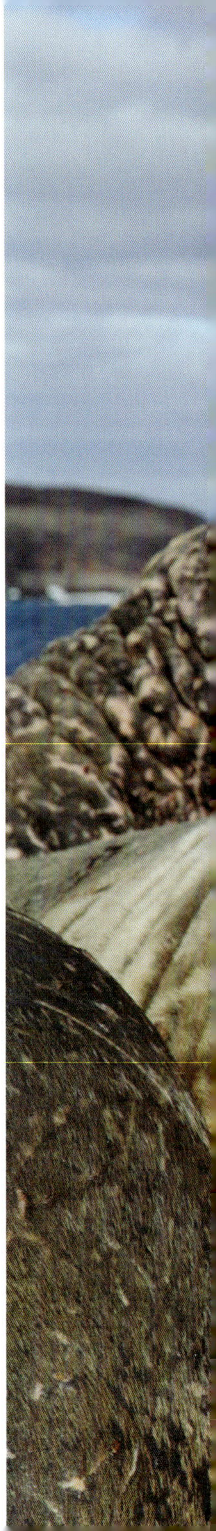

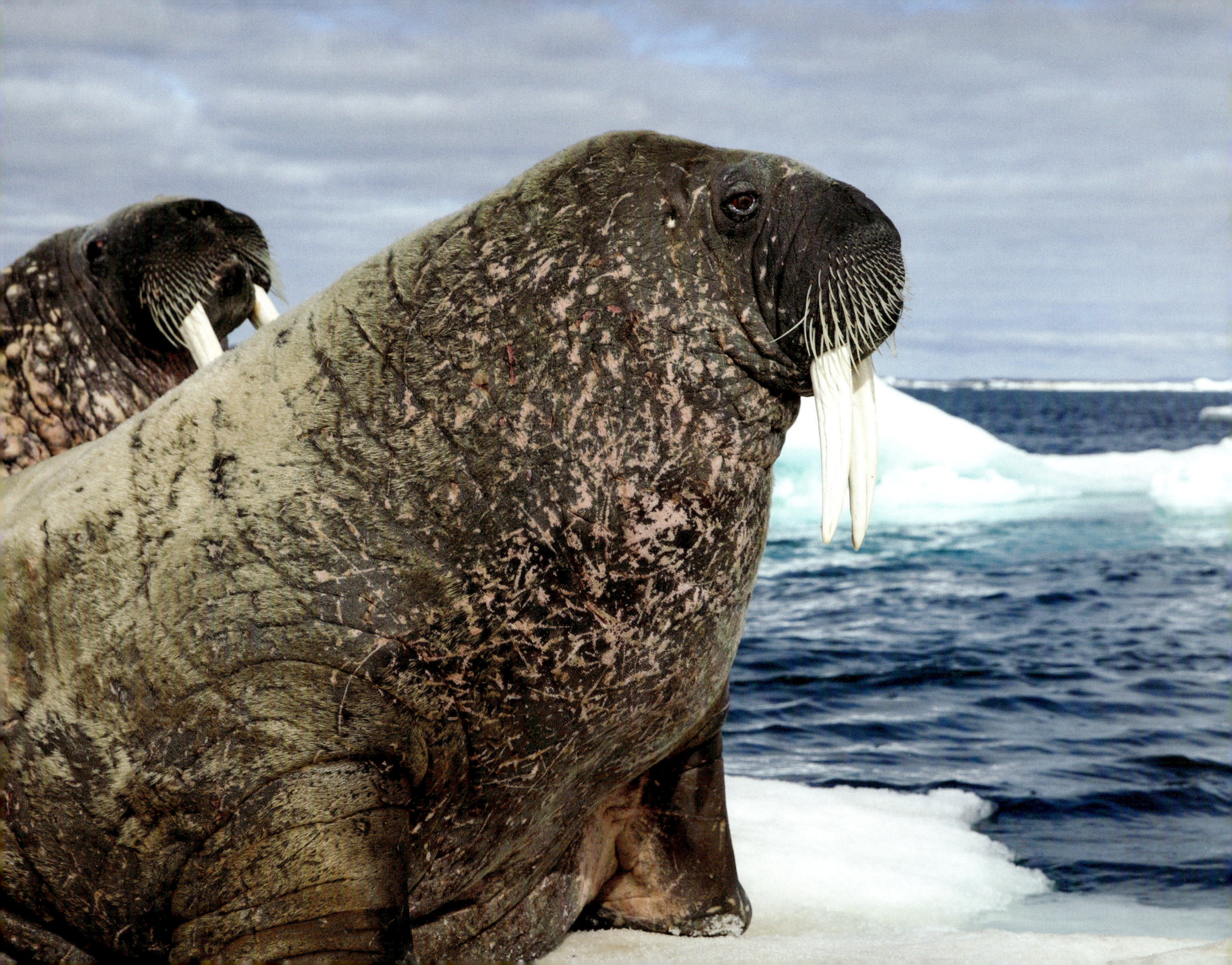

The tusks of male walruses are generally thicker and longer than those of females. The largest tusks have been recorded at up to one metre long and weigh up to five kilograms each. The front of the skull of a walrus is much stronger and thicker than related species, such as seals, in order to support these heavy tusks. The tusks of walruses are modified upper canine teeth that are elongated. Walrus tusks are used for many purposes, such as hauling out on ice, breaking ice, aggressive encounters with other walruses, and defending young, and are also thought to be used for digging shellfish on the bottom of the ocean.

*— Jordan Hoffman, Iqaluit*

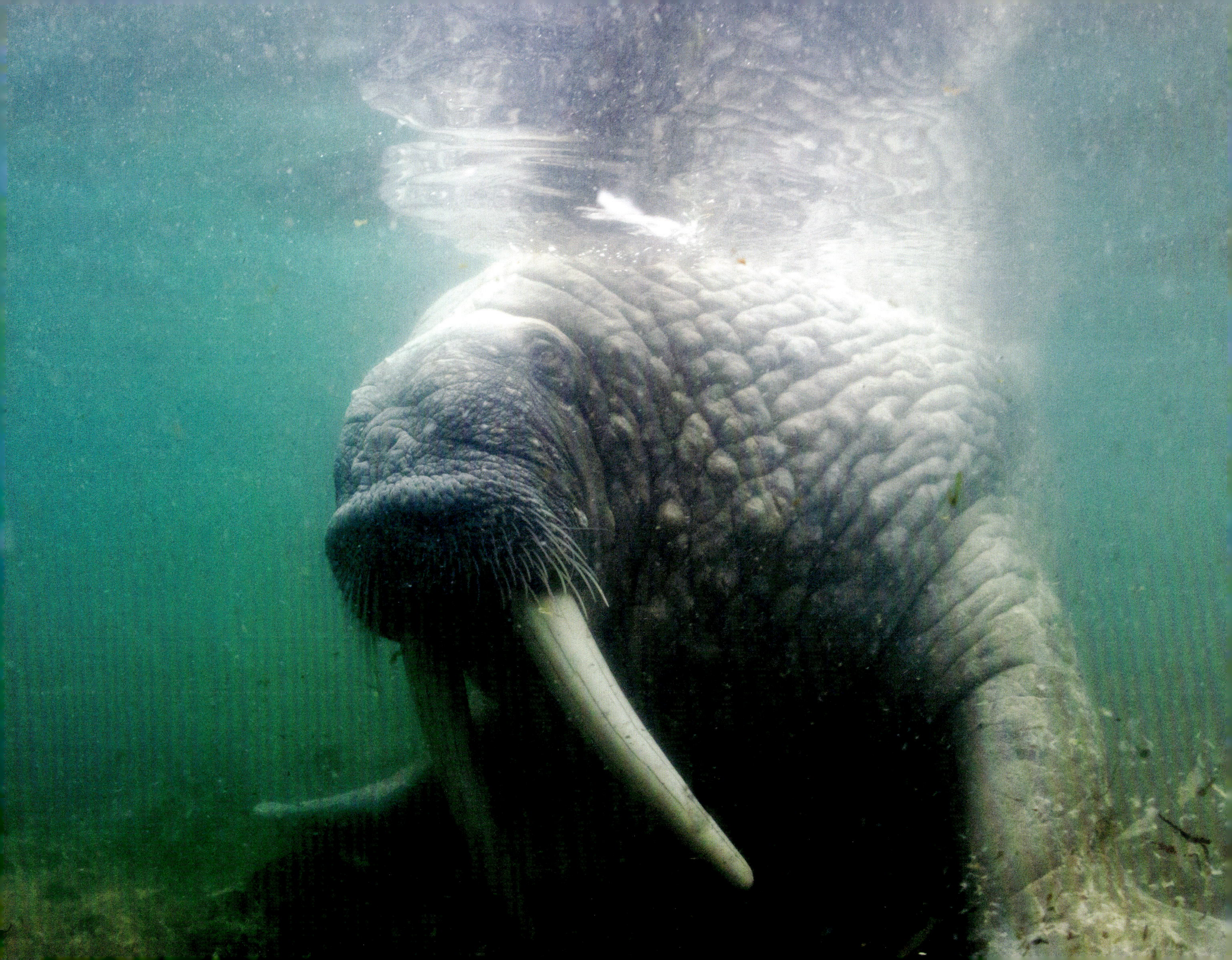

The hike from the seabird camp to the walrus haulout was eight kilometres over rocky and varied terrain. Unlike the rest of Coats Island, which is mostly flat and covered with wetlands and muskeg, the north side of the island is rocky, reaching a maximum height of 185 metres above sea level. The exposed rocky cliffs that rise from the sea provide ideal habitat for nesting seabirds like the thick-billed murre, the bird I was on the island to study. This was the second time I had done the long hike to the haulout beach. The first time was on Canada Day, our first day off since arriving on the island in mid-June. I had been so excited to see the incredibly odd-looking creature that I had only ever seen in textbooks. I was motivated by the stories from other researchers who had worked at the camp in previous years, who said that the walrus haulout was something special. Groups of eight hundred individuals were common, with the rocky peninsula and sandy beach both littered with these one-ton moustached animals. Without fully realizing it, we had been too early; the walruses had not yet arrived in the waters around the island. It was the beginning of August when I returned the second time, the last full day on the island before the plane arrived to take us back to Iqaluit. My last chance.

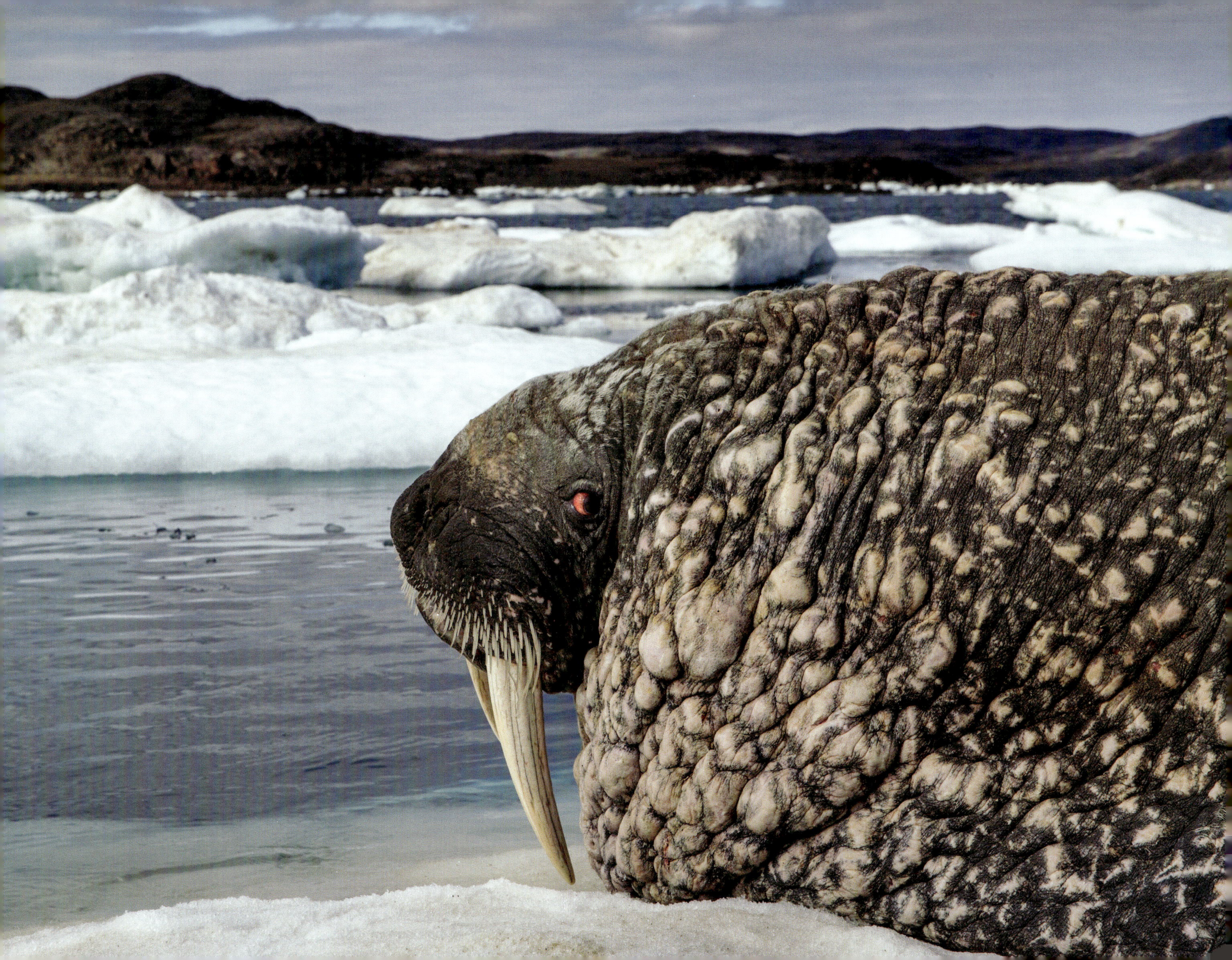

As we came into view of the beach, my heart sank again. No walruses. Dejected, but wanting to make the most of the trip, we walked out onto the peninsula to see if we could find any remnants or clues that the walruses had been there. As I looked out over the other side of the peninsula towards the other beach, I saw them: two walrus heads poking up from the water. They were maybe three hundred metres offshore and headed towards us! I called the other researchers who had come, and we sat and waited to see if they would come closer. Sure enough, within five minutes they were within four metres of shore and the two had materialized into a small pod of twelve individuals. They seemed to be as curious about us as we were about them, surfacing and trying to get more of their bodies out of the water to get a better look at us. It was then that we could see their huge tusks and get an idea of their enormity. It is hard to believe, but they were shy, always diving under the waves when we made a move to get a better view of them. They never came to land but stayed within four to twenty metres of the shoreline. I caught sight of a baby walrus with its mother, bobbing vertically in the water. The mother walrus appeared to be holding the baby

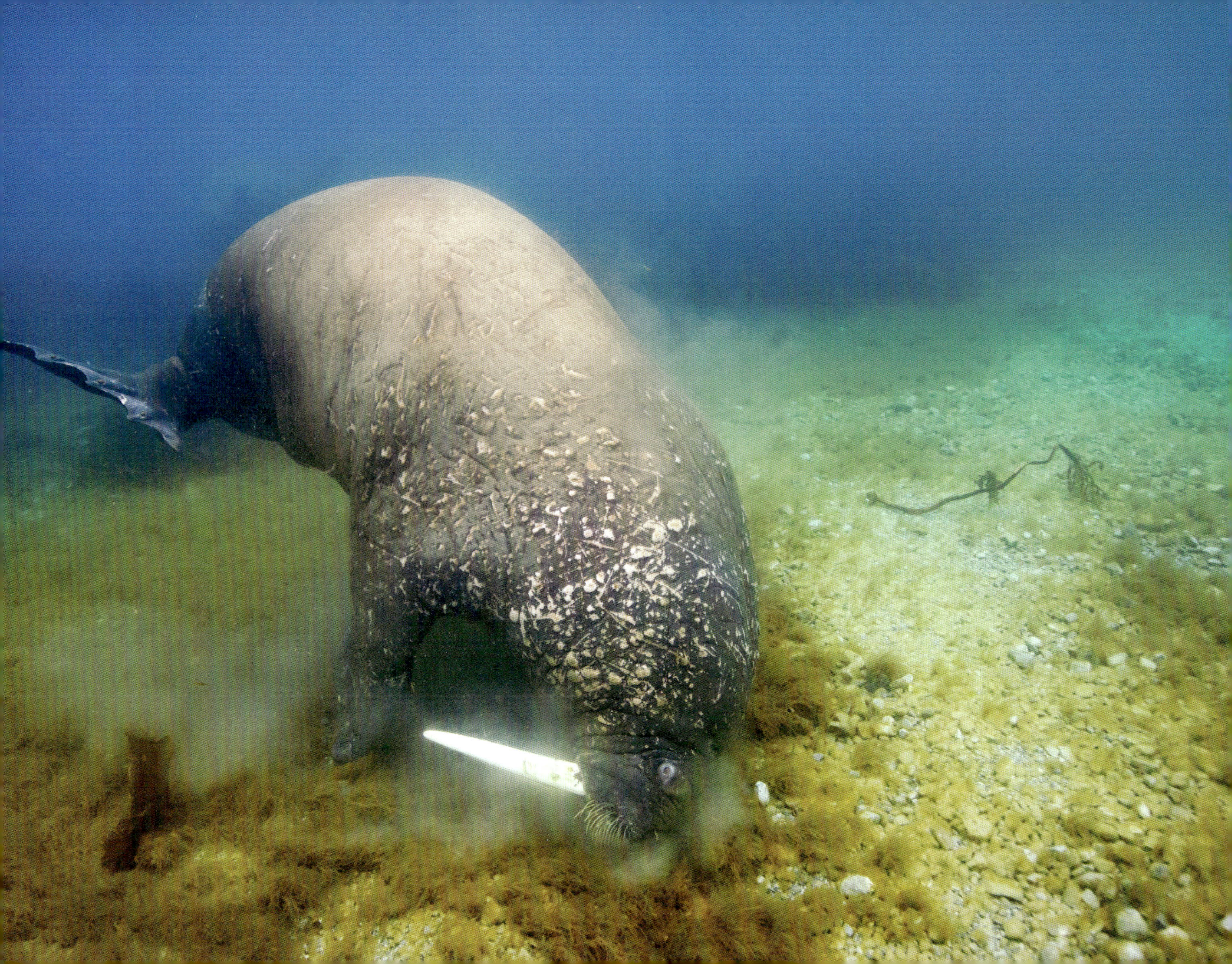

walrus up with its flippers to help the baby's head stay above the waves, or maybe to allow it to rest. We must have watched the pod for a good hour before heading back up the rocky terrain in the direction of our camp.

*—Sarah Poole, Coats Island*

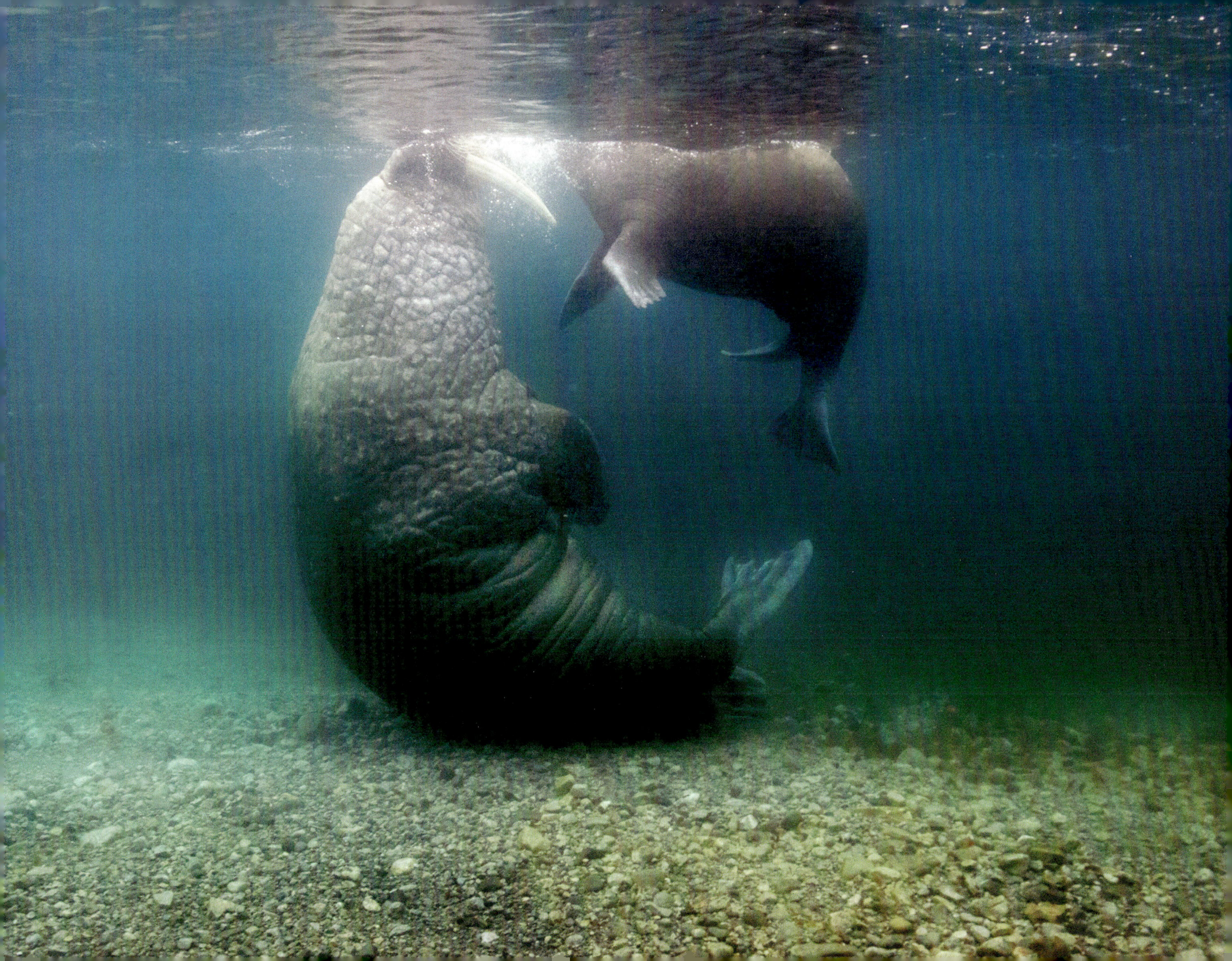

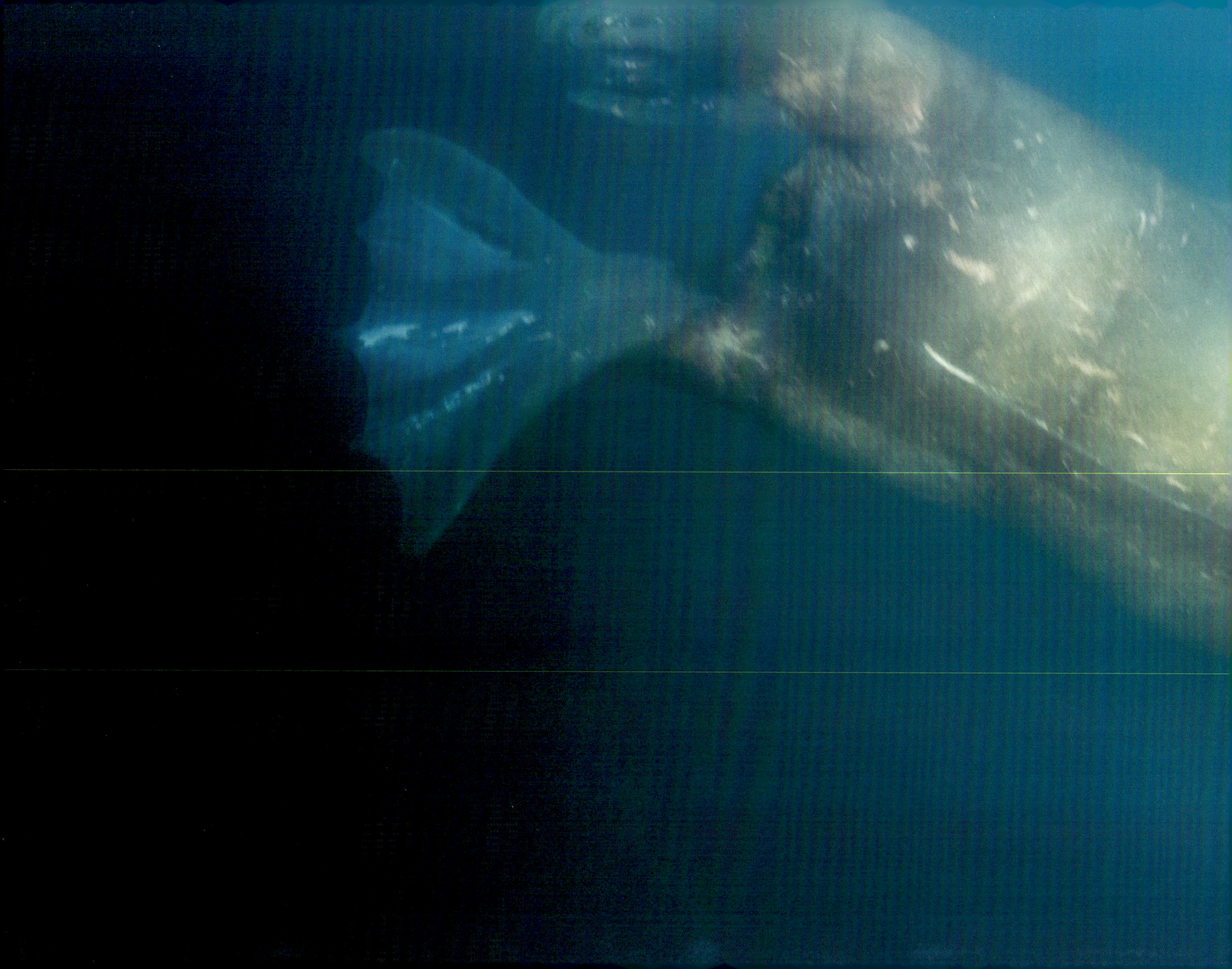

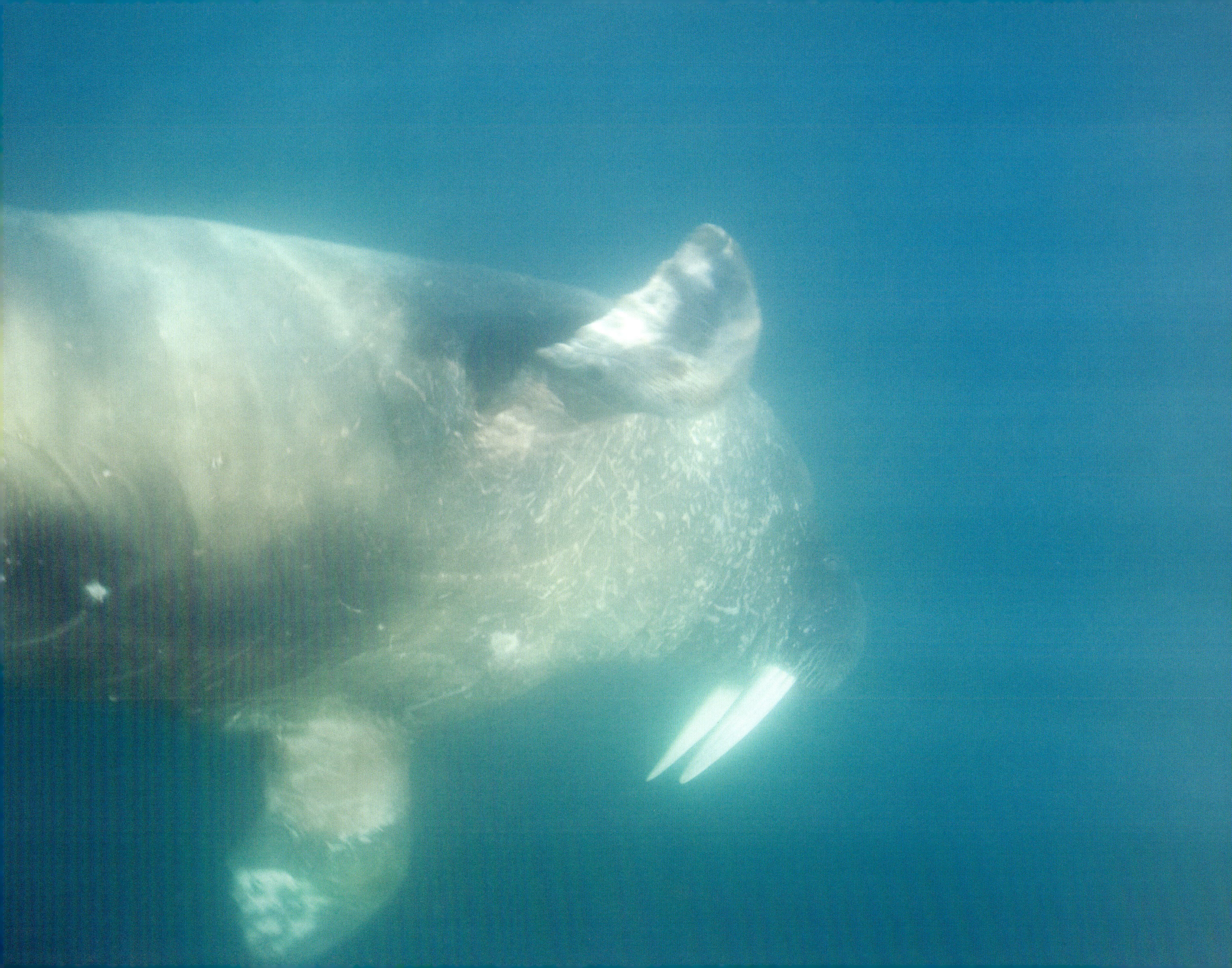

In olden times, the muskox and the walrus were great friends. The muskox gave his horns to the walrus, and the walrus gave the muskox his tusks. But the muskox found that the tusks cracked in cold weather, so that he could not hear anything but the noise of the cracking tusks. Therefore, they gave back to each other the tusks and horns in the spring.

*—Traditional story recorded on the west coast of Hudson Bay, 1800s*

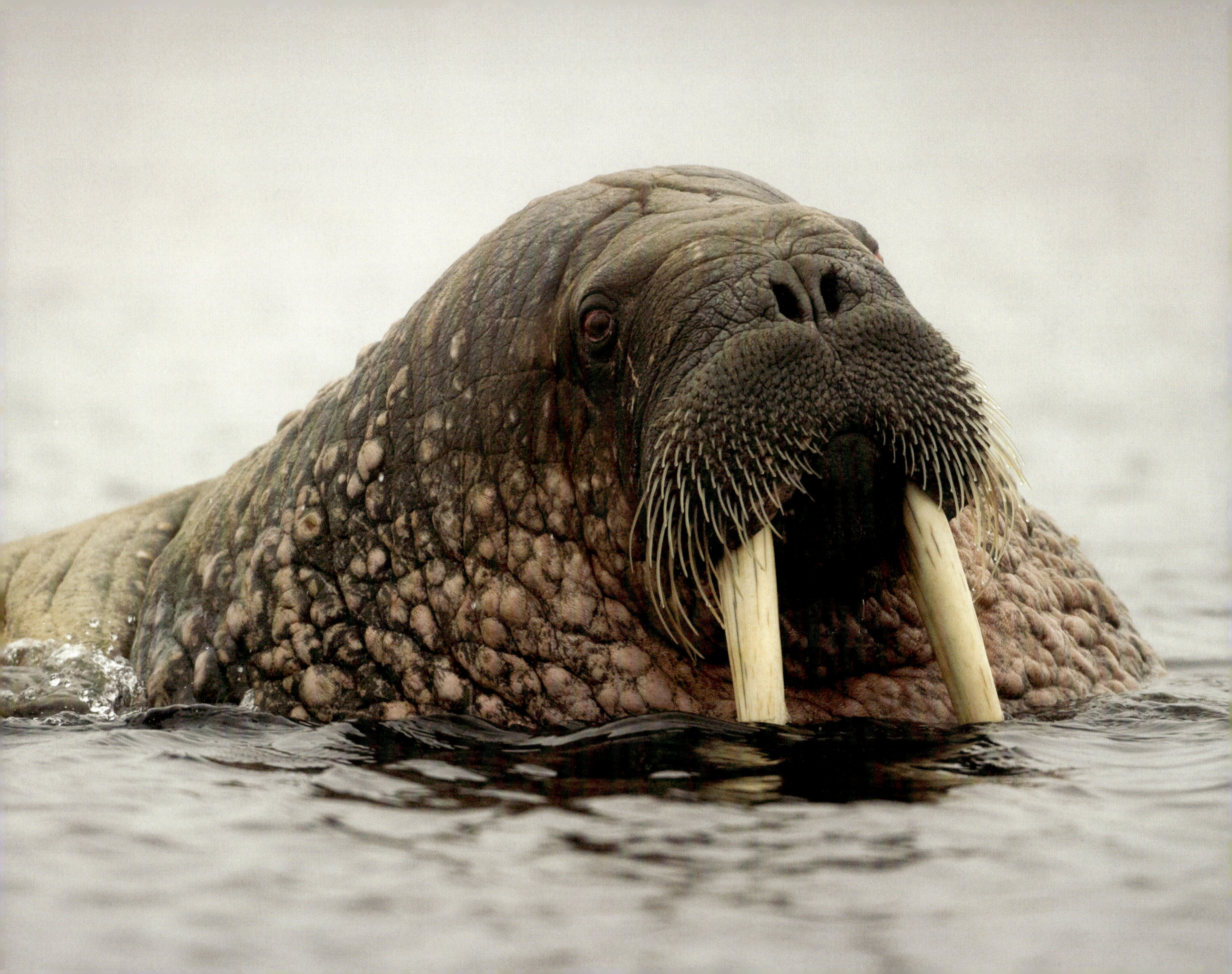

A long time ago, a woman transformed her old sealskin jacket into a walrus. She put antlers on its head, and then put it into the water. It looked very good. Then she transformed her trousers into a caribou. The black part became the back of the animal, while the white part became its belly. The waistband was used to make its legs, and the connecting parts of the trousers were used to make its loins. Then she put tusks in its mouth. It looked very good, and she set it free.

When the caribou saw a man, it went up to him and killed him with its tusks. Then the woman called both the walrus and the caribou to come to her. She took the tusks out of the caribou's mouth and put them into the walrus's head, and she took the antlers from the walrus and put them on the caribou's head. She also took out some of the caribou's teeth, and she kicked its forehead so that it became flat, and so that the eyes protruded.

In this way she punished the caribou for having killed the man. Then she said to the caribou, "You shall never come near the walrus. Stay far away inland." Ever since that time, whenever a caribou smells a man, it is afraid.

*—Traditional story recorded in the Cumberland Sound area, 1800s*

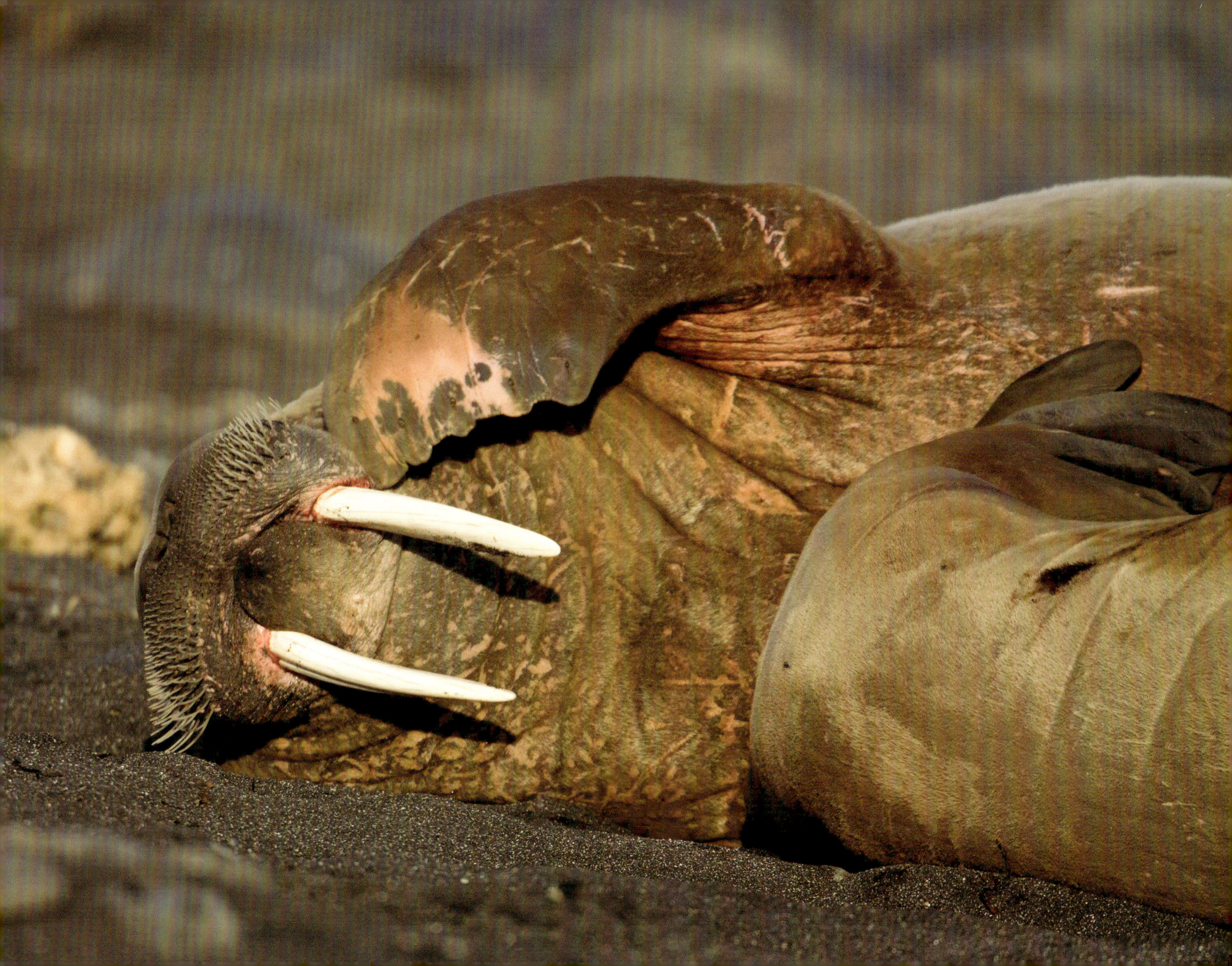

When I first started going walrus hunting with other hunters, I was told not to hunt walruses that were alone, because they're aggressive. When I went boating with a friend, we saw a walrus that was alone, and the walrus followed us for about two hours . . . . My friend was driving the boat and he told me that I should never approach the walrus because it is aggressive—we'd have to escape from the walrus.

The walrus dove down, and it disappeared for a while. Then it surfaced right next to the boat and tried to attack the boat. We turned and drove away from it. It followed us for a while. We shot near it to scare it away, but it wasn't even scared of the shots. We ended up driving the boat around, turning until we finally got away from it.

*—Sandy Irngaut, Iglulik*

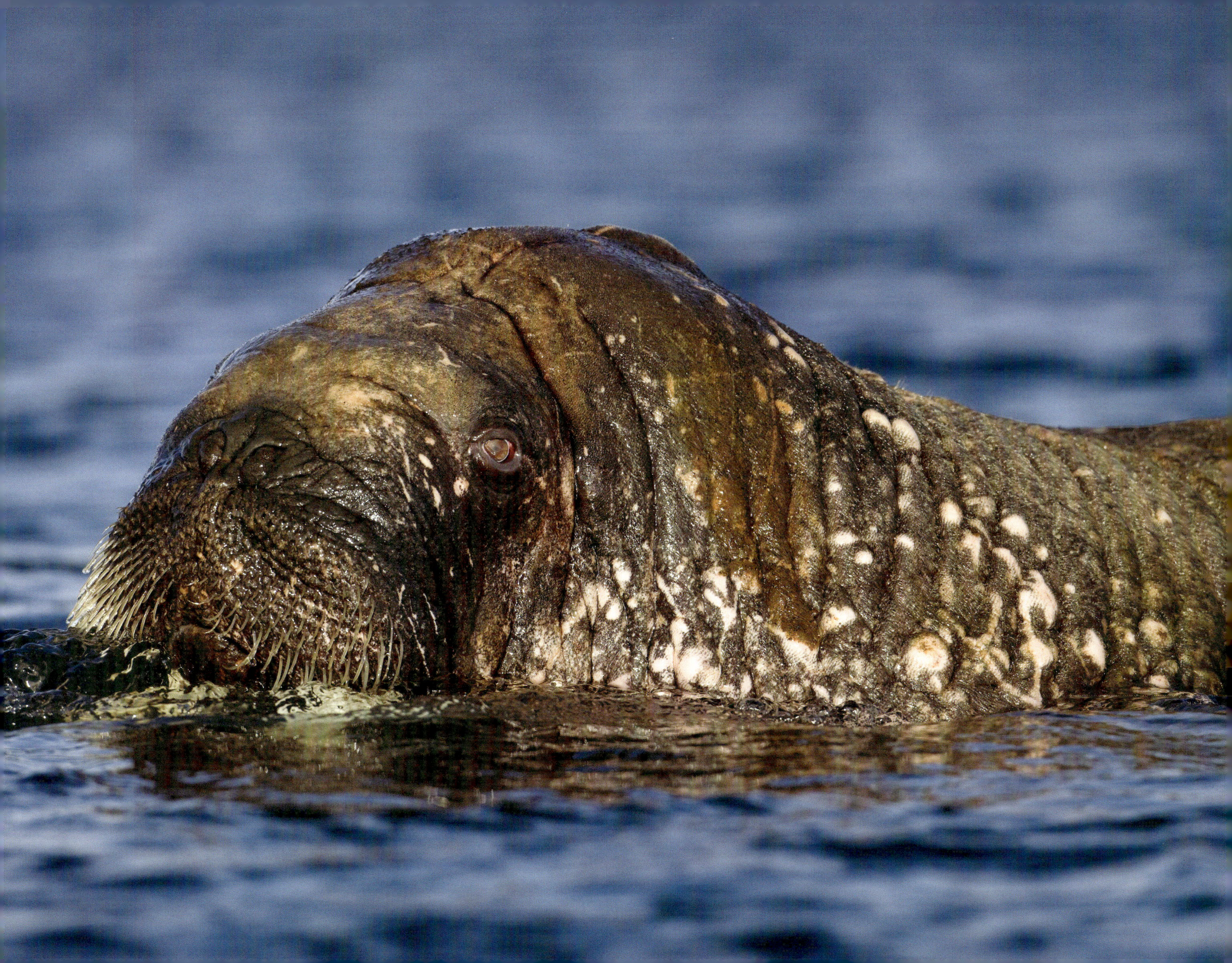

When I was boating, I saw a walrus with long tusks. It was an aggressive one. It looked like it was going to surface far from my boat, but it surfaced close to the boat. I drove away each time I saw it surface, but it was following me.

*—Samueli Ammaaq, Iglulik*

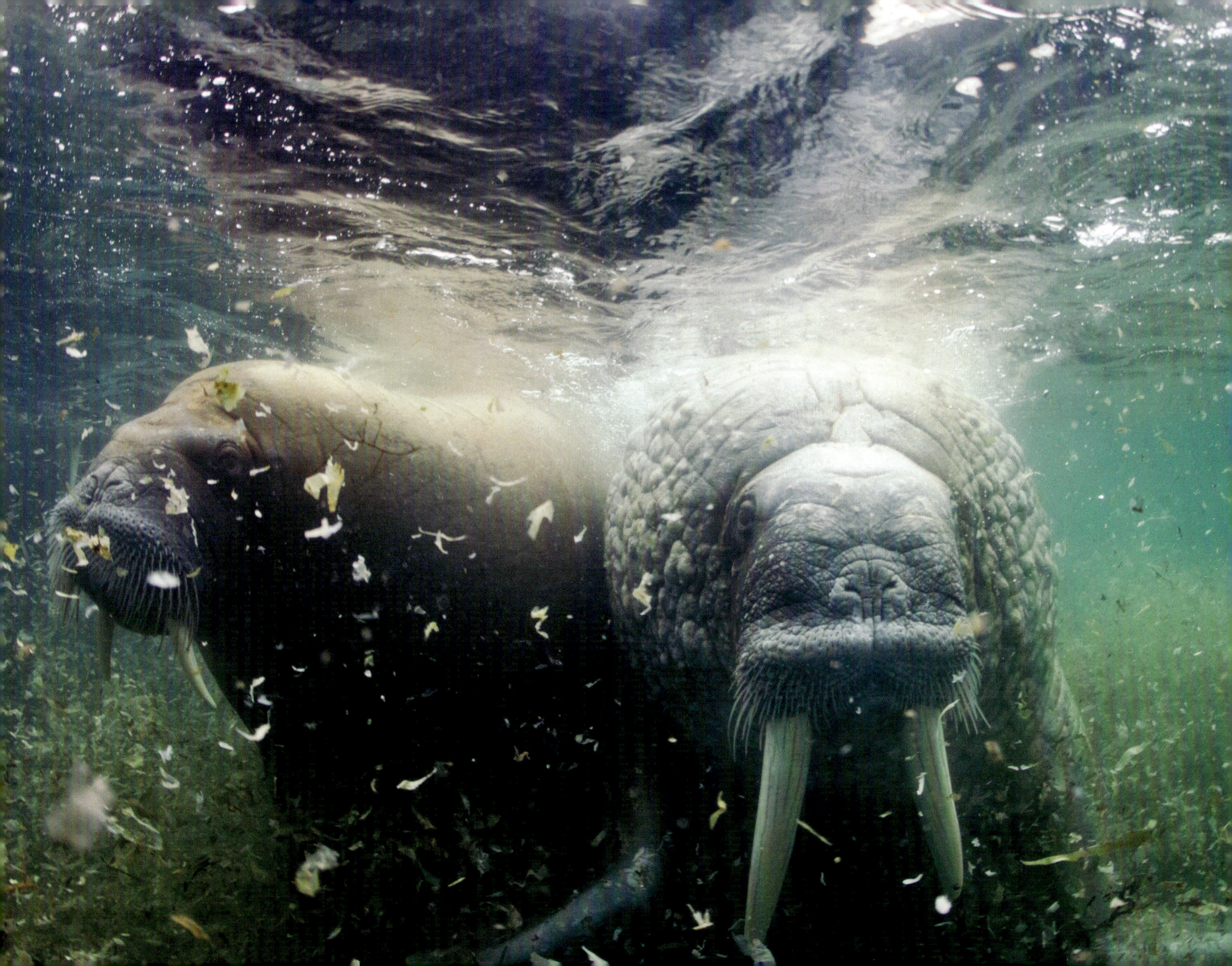

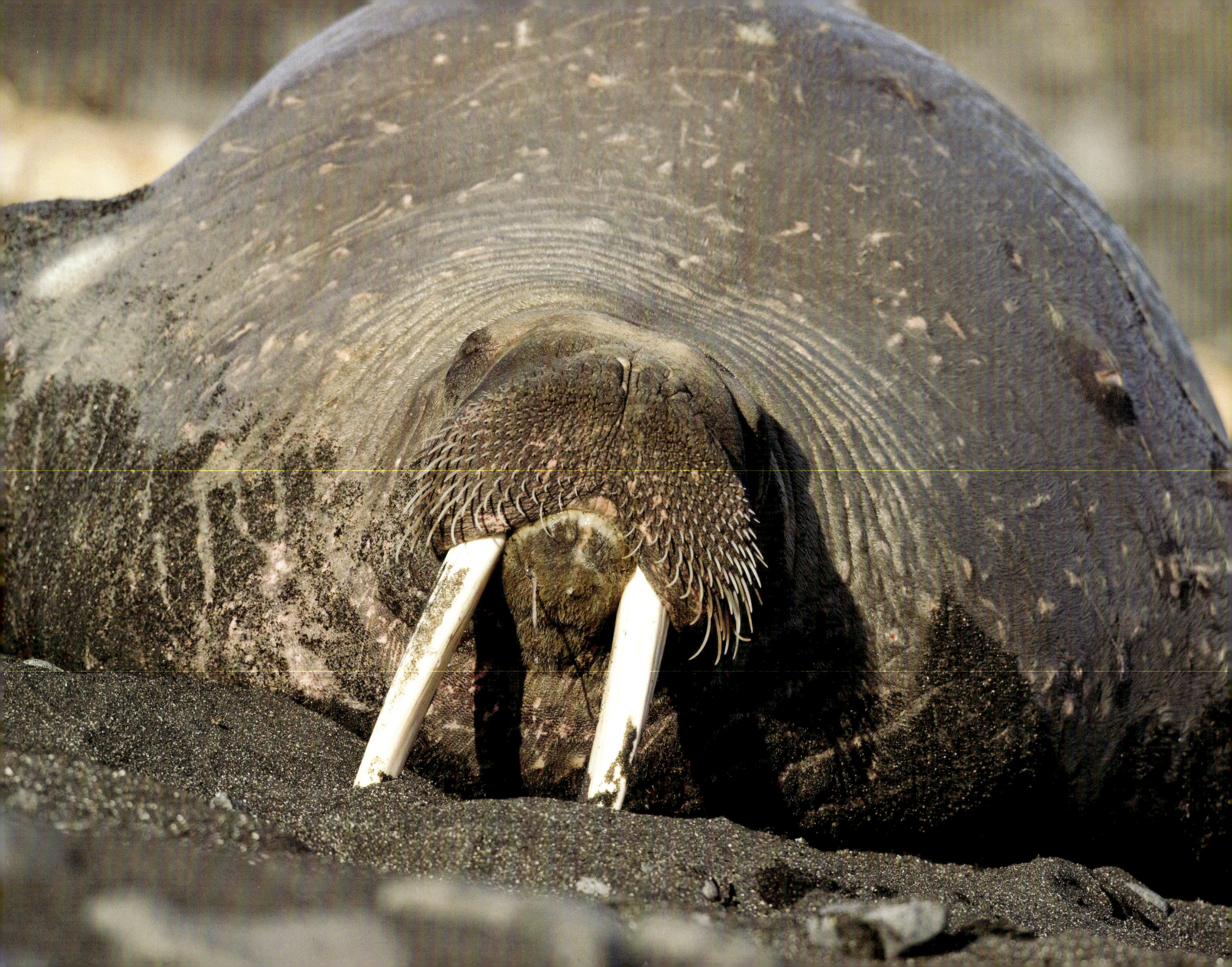

I've come very close to a sleeping walrus while I was on the ice. I saw a walrus sleeping on the ice while I was boating. It didn't seem to hear the boat motors, even when I was talking, but when someone made a high-pitched noise, it got up and jumped toward us . . . . I rushed to get out of the way and hit a piece of ice. The ice punctured my boat, but luckily, I was able to make it home with a hole in my boat. We weren't trying to toy with the walrus, we just wanted to see if it was dead.

*—Samueli Ammaaq, Iglulik*

Usually walruses eat clams, mussels, snails, and other bottom-dwelling creatures, but sometimes they eat seals. Mainly the ones that are living in deep water, coming from Arctic Bay area or farther, are more fond of seals . . . . Young people do not know about walruses. If [the walrus's] body is showing too much, while the head is underwater, that means it is dangerous. The smaller ones are usually more dangerous, and big males that are alone, like any other animal.

*—Herve Paniaq, Iglulik*

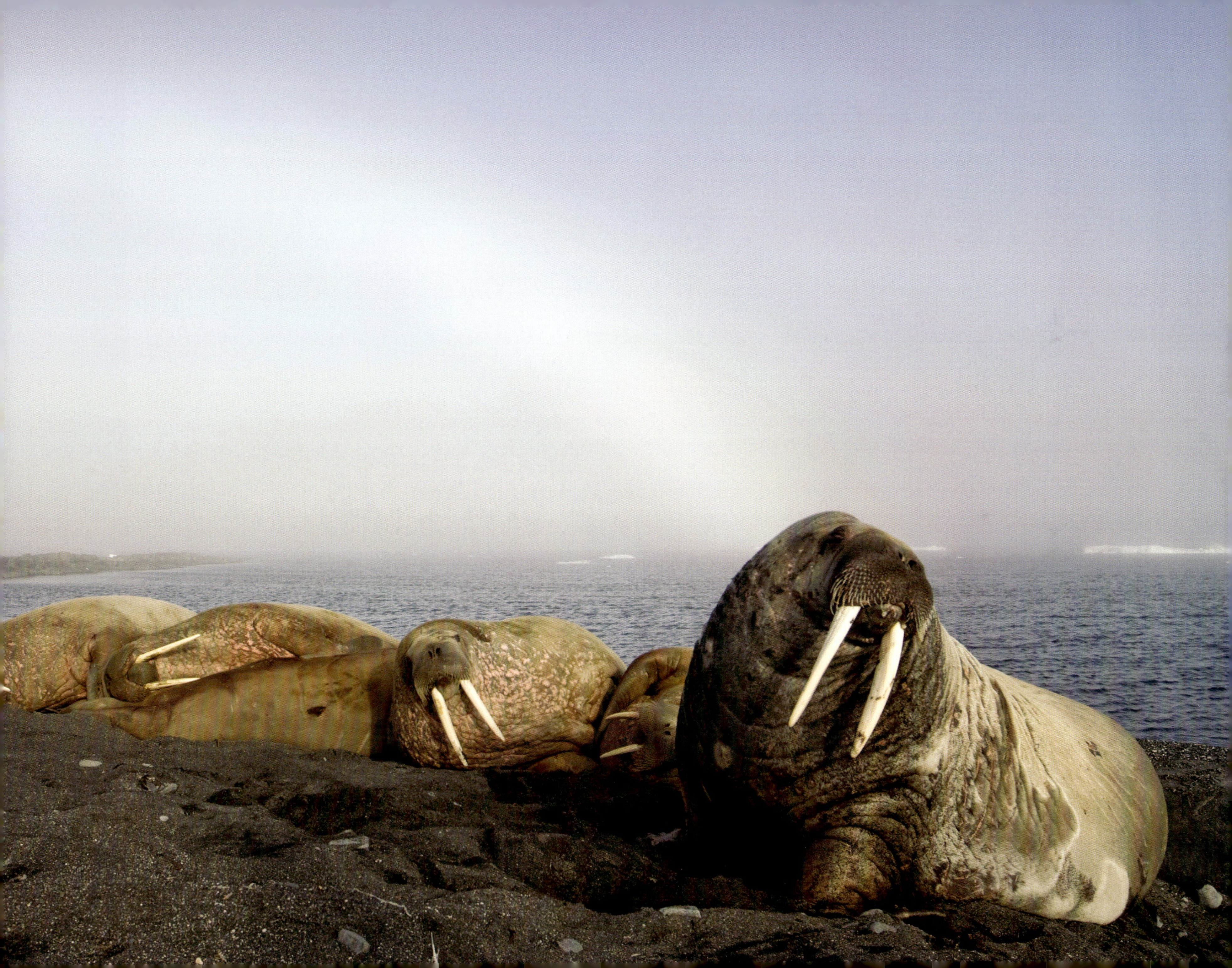

I could not sleep,
for the sea lay so smooth
near at hand.
So I rowed out,
and a walrus came up
close beside my kayak.
It was too near to throw,
and I thrust the harpoon into its side,
and the hunting float bounded over the water.
But it kept coming up again
and set its flippers angrily,
like elbows on the surface of the water,
trying to tear the hunting float to pieces.
In vain it spent its utmost strength,
for the skin of an unborn lemming
was sewn inside as a guardian amulet,
and when it drew back, blowing viciously,
to gather strength again,
I rowed up and stabbed it
with my lance.
And this I sing
because the men who dwell
south and north of us here
fill their breathing with self-praise.

—*Walrus hunting song recorded in the Iglulik area, 1920s*

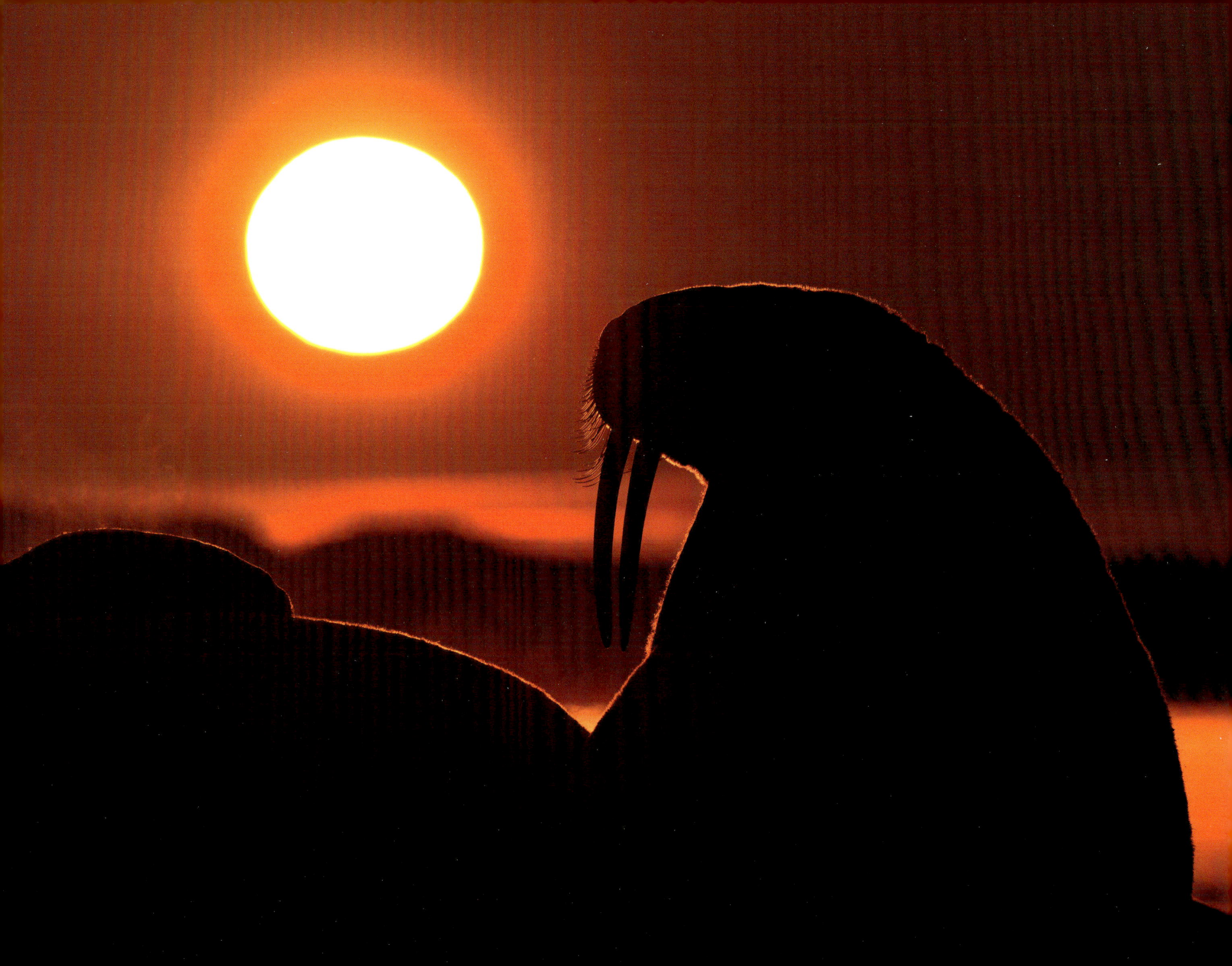

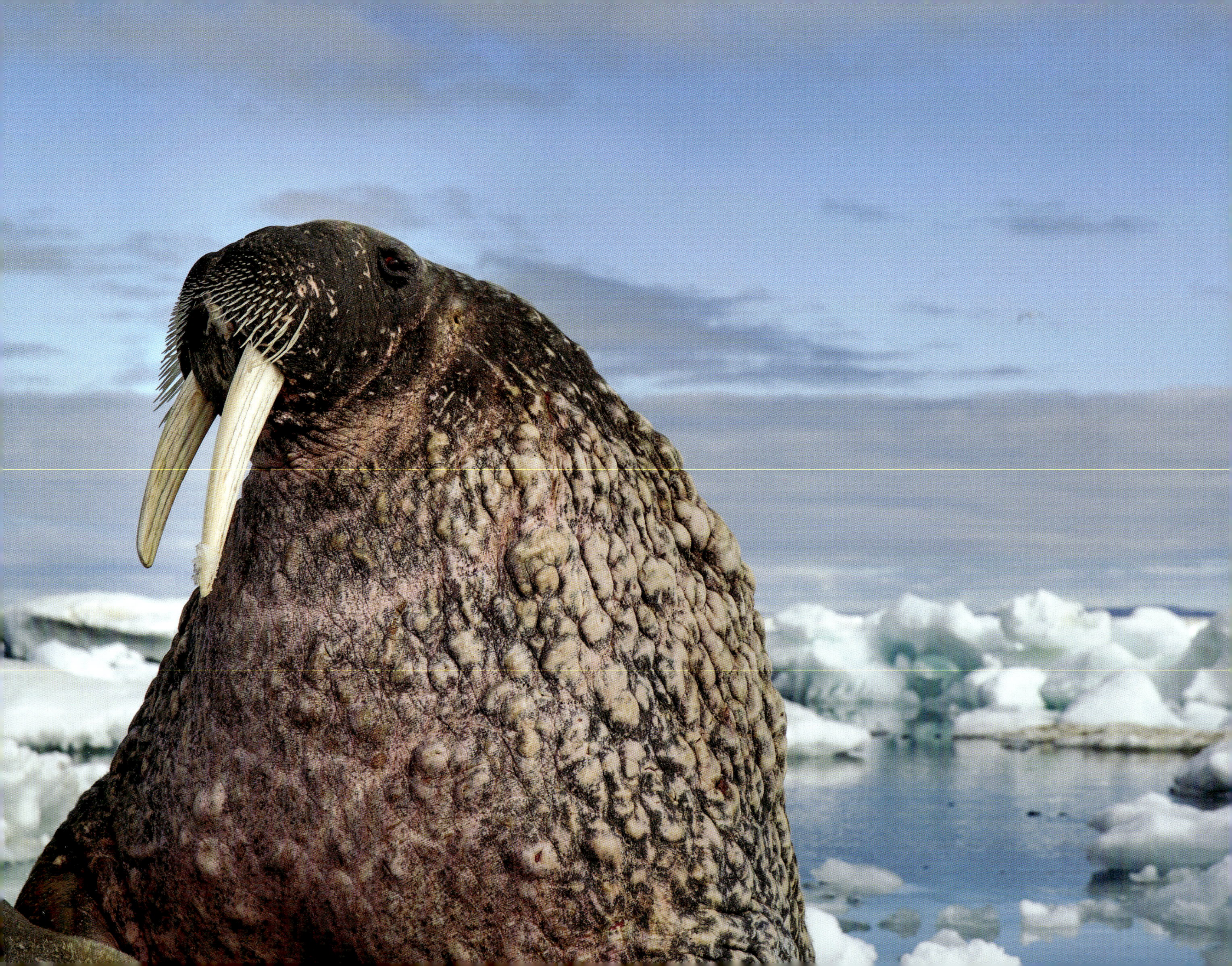

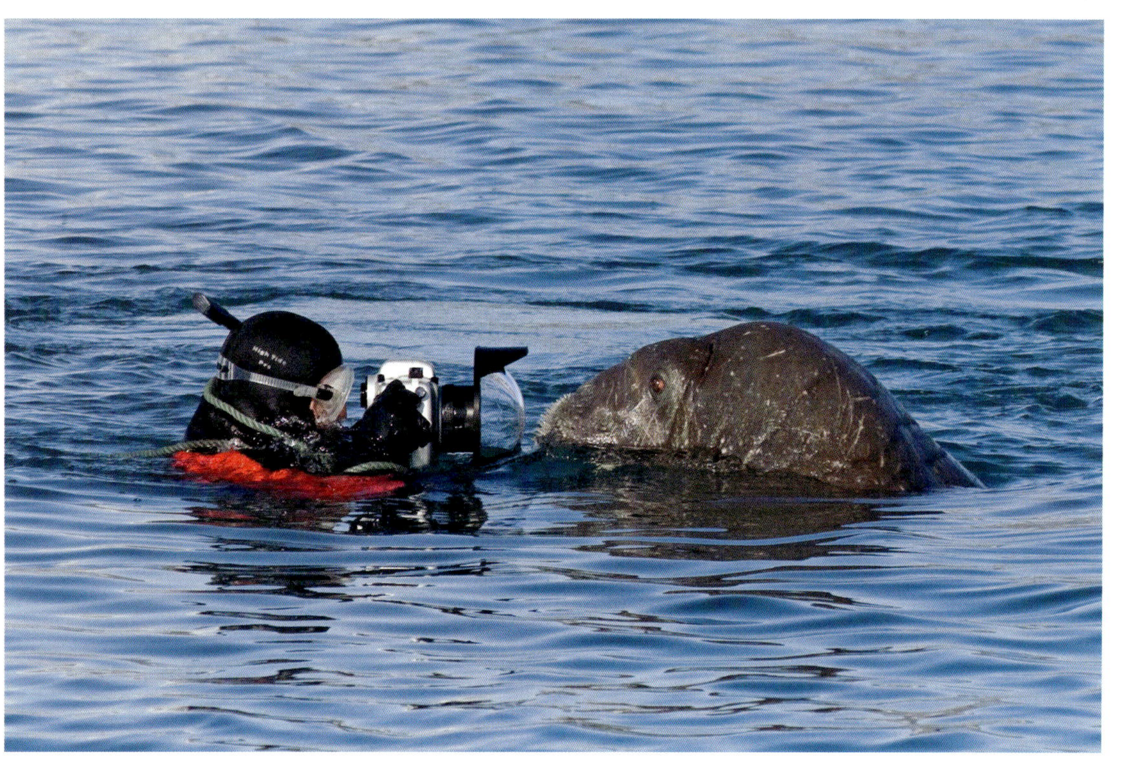

## Paul Souders

Paul Souders is a professional photographer who has been travelling around the world, across all seven continents, for more than thirty years. His images have appeared around the globe in a wide variety of publications, including *National Geographic*, *Geo* in France and Germany, and *Time* and *Life* magazines, as well as hundreds of publishing and advertising projects. His recent photography work in the Arctic has drawn wide acclaim, including first-place awards at the BBC Wildlife Photographer of the Year competition in 2011 and 2013, the *National Geographic* Photo of the Year contest in 2013, and Grand Prize in the 2014 Big Picture Competition. Over the last three decades he has visited more than sixty-five countries and has been slapped by penguins, head-butted by walruses, terrorized by lions, and menaced by vertebrates large and small.

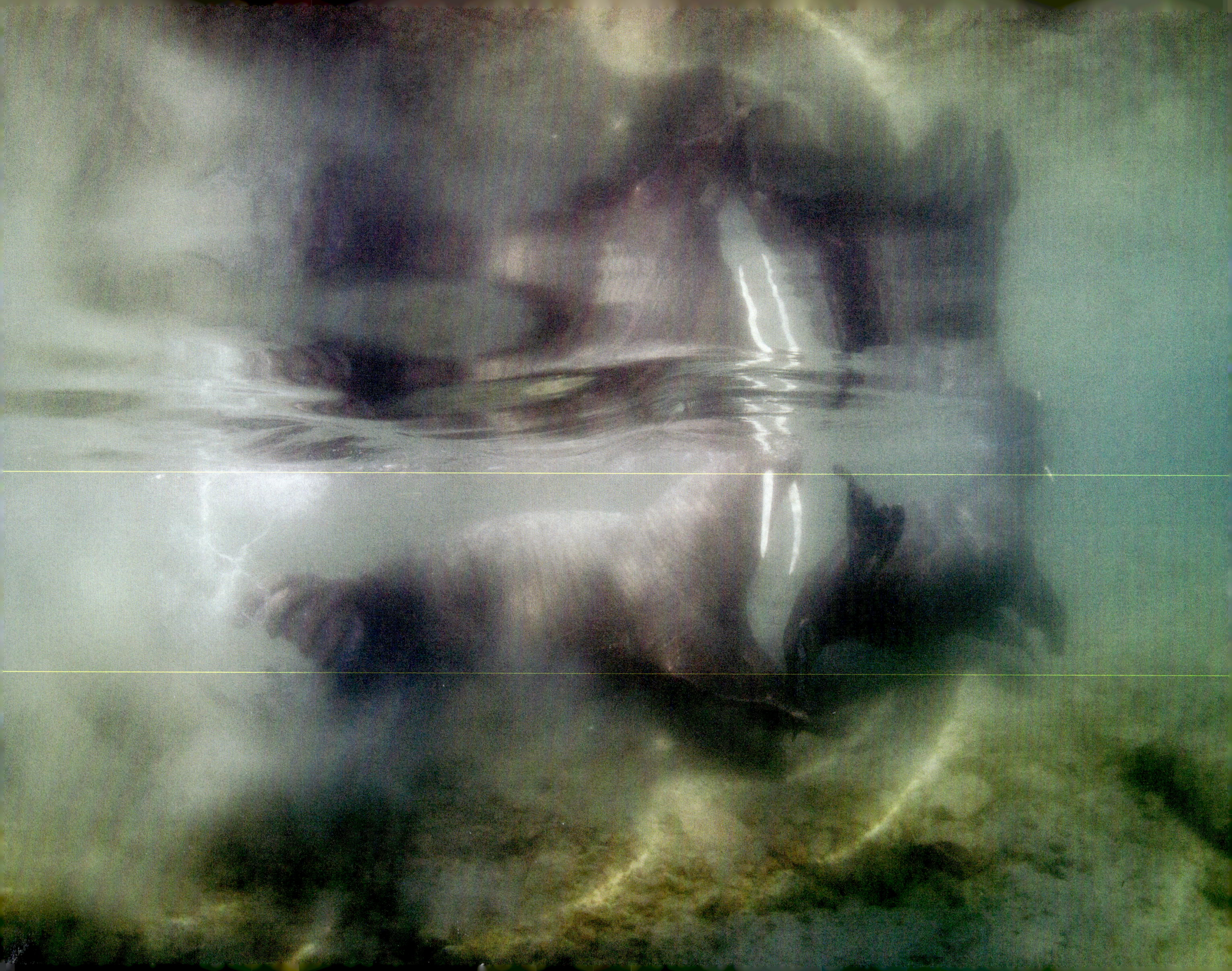